T-Mobile G1™ For Dummies®

G000122719

Application   Touchscreen   Signal strength   Appli

                     Data transfer   Battery meter

myFaves

2:47 PM

Dialer   Contacts   Browser   Maps

End

Back

MENU   Trackball

Home

Send

Search

Gmail Key Combinations

From the Inbox view (or any other label):

Action	Key Combination
Compose new e-mail	Menu+C
Refresh or synchronize	Menu+U

From an individual e-mail:

Action	Key Combination
Reply to all	Menu+A (while looking at an e-mail)
Reply	Menu+R (while looking at an e-mail)
Forward	Menu+F (while looking at an e-mail)

T-Mobile G1™ For Dummies®

Cheat Sheet

Quick Launch Keys

From anywhere in the G1, you can press the Search key on the keyboard plus a letter to launch different applications. Several of these quick launch keys have been assigned by default, but you can change the defaults and add more from the Settings application.

Application	Key Combination
Browser	+b
Contacts	+c
Email	+e
Gmail	+g
IM	+i
Calendar	+l
Maps	+m
Music	+p
Messaging	+s
YouTube	+y

Accented Characters

By holding down certain letters on the keyboard, you can get a list of available accented characters that can be produced from that letter. Touch any of the shown accented letters to insert it into your text.

Letter	Available Accents
A	À, Á, Â, Ä, Æ, Ã, Å
C	Ç
E	È, É, Ê, Ë
I	Ì, Í, Î, Ï
N	Ñ
O	Ø, Œ, Õ, Ò, Ó, Ô, Ö
S	§, ß
U	Ù, Ú, Û, Ü
Y	Ý, Ÿ

Browser Key Combinations

Action	Key Combination
Go to Bookmarks	Menu+B
Go to History	Menu+H
Go forward	Menu+K
Refresh	Menu+R
View windows	Menu+T
Close this window	Menu+W

Copyright © 2009 Wiley Publishing, Inc. All rights reserved. Item 9340-6.

For more information about Wiley Publishing, call 1-800-762-2974.

For Dummies: Bestselling Book Series for Beginners

by Chris Ziegler

Wiley Publishing, Inc.

T-Mobile G1™ For Dummies®

Published by
Wiley Publishing, Inc.
111 River Street
Hoboken, NJ 07030-5774

www.wiley.com

WILEY

About the Author

Chris Ziegler is a mobile enthusiast who can rarely be spotted using the same phone twice. Although his passions include driving, flying, motorcycling, and pretty much anything that involves gasoline, he's usually piloting nothing more than his own computer as Associate Mobile Editor for the technology blog Engadget. A native of Michigan, he currently lives in Chicago with two Macs, two PCs, and a pile of portable gizmos taller than he.

Dedication

To my fantastic parents, Brenda and Zig, for being far more patient and understanding than a son could ever ask a mother and father to be. I'd also like to thank Ryan Block for taking a chance on an unknown who'd never written a line of prose in his life; he put my career on a wild new path.

Acknowledgments

A world of thanks goes to Kyle Looper, my acquisitions editor, for reaching out to Engadget to make this book happen; to Susan Pink, my editor, for reminding me that readers don't always know what I'm talking about when I refer to it (and for so much more); to Andrew Hilsher, my technical editor, for making sure I know what the heck I'm talking about; to Peter Rojas for connecting me with Kyle; and to everyone at Wiley Publishing who's helped to put *T-Mobile G1 For Dummies* on the bookshelf. It's been quite a ride.

Publisher's Acknowledgments

We're proud of this book; please send us your comments through our online registration form located at `http://dummies.custhelp.com`. For other comments, please contact our Customer Care Department within the U.S. at 877-762-2974, outside the U.S. at 317-572-3993, or fax 317-572-4002.

Some of the people who helped bring this book to market include the following:

Acquisitions and Editorial

Project Editor: Susan Pink

Acquisitions Editor: Kyle Looper

Copy Editor: Susan Pink

Technical Editor: Andrew Hilsher

Editorial Manager: Jodi Jensen

Media Development Assistant Producers: Angela Denny, Josh Frank, Kit Malone, and Shawn Patrick

Editorial Assistant: Amanda Foxworth

Sr. Editorial Assistant: Cherie Case

Cartoons: Rich Tennant (`www.the5thwave.com`)

Composition Services

Project Coordinator: Erin Smith

Layout and Graphics: Reuben W. Davis, Ronald Terry, Christine Williams

Proofreaders: Context Editorial Services, Amanda Graham

Indexer: Potomac Indexing, LLC.

Special Help: Ronald Terry

Publishing and Editorial for Technology Dummies

    Richard Swadley, Vice President and Executive Group Publisher

    Andy Cummings, Vice President and Publisher

    Mary Bednarek, Executive Acquisitions Director

    Mary C. Corder, Editorial Director

Publishing for Consumer Dummies

    Diane Graves Steele, Vice President and Publisher

Composition Services

    Gerry Fahey, Vice President of Production Services

    Debbie Stailey, Director of Composition Services

Contents at a Glance

Table of Contents

Part IV: Adding Software and Media 255

Chapter 16: Buying Music with Amazon MP3 257

Chapter 17: Filling Your Phone with Media 269

Chapter 18: The Android Market 277

Introduction

· ·

I'd like to take this opportunity to welcome you to *T-Mobile G1 For Dummies*. Regardless of whether you have recently purchased a G1 and need help getting started, are a seasoned professional looking for a quick refresher, are considering purchasing a G1 and would like to get a feel for what the device is all about, or just need to know what a G1 *is*, you've come to the right place.

In a nutshell, the G1 is a state-of-the-art cellphone developed by Google, HTC, and T-Mobile that features an expansive touch screen, a full keyboard, navigation capabilities, and a whole lot more. It connects to both Wi-Fi networks (like those found in homes and offices) and also to T-Mobile's own high-speed data network to offer you speedy access to e-mail and the Web no matter where you happen to be.

Of course, a phone as amazing and feature-packed as the G1 is far more than just a phone — it's an administrative assistant, a PC away from your PC, a movie theater, a jukebox, a mobile office, a gaming partner, and pretty much everything in between. This book shows you how to integrate the G1 into every part of your life (or only some parts, if you prefer!) and how to get the most out of it every step of the way.

The G1 is also the very first phone to be released using Google's Android software, which is a standard operating system that many phones from different manufacturers and wireless carriers will use. This approach has several notable advantages, but most importantly, it means you'll be able to take much of the knowledge you discover throughout your journey with the G1 with you should you move to other Android devices down the road. Standardization: Anyone who's migrated between several BlackBerrys, Windows Mobile devices, or even Windows-based PCs will be able to appreciate the benefit of that!

I cover the fundamentals of Android and the G1, but I also show you great tips and tricks along the way. Read the book cover to cover if you like, or read just the sections that interest you most — either way, you can expect to find some cool tidbits that make the G1 an even handier gadget than you thought it was.

About This Book

T-Mobile G1 For Dummies can be read from beginning to end, either with or without a G1 in front of you. If you happen to have a G1 handy, that's great because you'll be able to follow along with the many examples. But if you don't have a G1, you'll end up with a great idea of how it works for both simple and more complex tasks. (You can think of this book as the ultimate buyer's guide!)

The book also makes a great reference that you can quickly consult time and time again. And because the G1 runs the standard Android platform, many of the book's concepts will remain useful if you decide to move on to another Android-based phone in the future. Who knows — you could be passing *T-Mobile G1 For Dummies* down to your great-grandkids someday.

Conventions Used in This Book

Here are some conventions I hope you'll find useful:

- ✔ The G1 has *buttons* below the screen. When I ask you to *push* a button, I am referring to one of these buttons on the phone.
- ✔ The G1's screen will show *icons* at times. If I want you to activate an icon, I ask you to *touch* it.
- ✔ You can *swipe* the G1's screen by moving your finger across it. I'll occasionally ask you to swipe to perform an action on the G1.
- ✔ *Flicking* is similar to swiping but your finger loses contact with the screen while your finger is still in motion. Flicking causes the G1's screen to continue scrolling for a bit — handy for rapidly moving through pages of information.
- ✔ The keyboard that comes with the phone has, well, *keys*. You *press* a key.

What You're Not to Read

If you've already powered on your G1, connected it to your Google account, and feel comfortable with the basics of navigating between screens and screen items, you can safely skip Chapters 1, 2, and 3. If you feel comfortable with call handling and text messaging, that means you're ready to dive into the G1's applications, and you can move right past Chapter 4.

Foolish Assumptions

Because you bought this book, I figure that you fall into one of three categories:

- ✔ **You've already purchased a G1.** Now you're looking to maximize your investment, find out a little more about what Android's all about, and pick up a few tips and tricks along the way. If so, congratulations — by choosing this book, you've made *two* smart choices!

- ✔ **You haven't taken the plunge yet but want to find out more about whether Android and the G1 are right for you.** I call the members of this group the "on-the-fencers." This book does an excellent job of spelling out Android's capabilities and how the G1 can fit into your life with its advanced e-mail capabilities, Web browsing, music support, and much more.

- ✔ **You don't know enough to recognize what Android is but do realize that what's good in your PC must be good in your pocket, too.** It turns out that you're absolutely right!

Regardless of what category you are in, you need an answer to a burning question: What does Android offer you? I'm going to try my best to get that answer to you in this book, and to start, it helps to understand where Android came from, what it is, and where it's going. I'm going to explore all that great stuff in Chapter 1. And don't worry — you'll be checking Gmail like a pro on your G1 in no time!

How This Book Is Organized

T-Mobile G1 For Dummies is organized into five themed parts, and each part is further subdivided into chapters covering individual topics.

Part 1: Introducing Android and the T-Mobile G1

Part I goes over the fundamentals by describing what Android is all about, why it's good for you, and how it ties into the G1. You find out about the G1's physical features, its capabilities, and how to make your first phone call. This is a great place to start if you're brand-new to the phone or trying to decide whether to take the plunge.

Part II: Putting the "Google" in Google Android

Google was deeply involved in the creation of the G1 and its software, and Part II goes over the fruit of that labor. All the things you love about Google on your desktop are in here: Gmail and Contacts, searching the Internet, Google Maps, Calendar, and more. Be careful, though — Google Maps on the G1 is a fun and *extremely* addicting pastime, so you could be spending a while on this part just for kicks!

Part III: Making the G1 a Part of Your Life

Part III shows you how the G1 can become a true mobile companion, putting music, pictures, and a great Web browsing experience right in your pocket. I also take a look at the IM application for keeping in touch with friends using instant messaging accounts, YouTube for accessing a vast library of entertaining videos on the go, and — for truly customizing the G1 and making it your own — the Settings application.

Part IV: Adding Software and Media

In Part IV, I show you how the G1 manages your data. You see how to add to your phone the music that's already have on your computer. In fact, the G1's music capabilities are so well-honed that it includes the Amazon MP3 program so you can purchase new tracks right on the device. (I show you how to use it, but keeping your spending in check is up to you!) I also get into Android's management of applications that you might install, how to delete them, and how to effortlessly get new applications through the Android Market.

Part V: The Part of Tens

Part V gets into some must-have apps, games, and accessories that'll make your G1 experience even cooler than it already is. I also throw in some sites you can visit to find out more about all the happenings in the fast-paced world of Android.

Icons Used in This Book

These are the time-savers and the eye-wideners. Watch for Tip icons to discover neat features, functions, and capabilities of your G1 that many users may not know.

When you put down *T-Mobile G1 For Dummies*, are you going to remember everything you've read verbatim? No, probably not. In fact, I'd be concerned if you did! These icons point out key takeaways that you'll want to put in your back pocket because you're likely to run into them down the road.

Take heed when you see one of these icons because the advice you find here could save you from some strife.

Where to Go from Here

Are you ready to read up on the most powerful phone you've ever seen? If so, let's move right into Chapter 1 without further ado, where you'll find out a little about the history of Android, what makes Android so cool, and why the future's looking bright for the T-Mobile G1.

Part I

Introducing Android and the T-Mobile G1

The 5th Wave   By Rich Tennant

"He seemed nice, but I could never connect with someone who had a ringtone like his."

In this part . . .

Generally speaking, the best way to master a smart-phone like the G1 isn't to dive right into the music player. (Well, maybe it's the best way if you're really into Depeche Mode and not so much into making phone calls, but I'm trying to cover the majority here.) I get things off on the right foot by talking a little bit about Android's past, present, and future in Chapter 1 and showing you the G1's physical features in Chapter 2. In Chapter 3, you find out about the Home screen (as the name implies, you'll see this screen a lot). I wrap things up in Chapter 4 with a thorough discussion of calling and text messaging.

Chapter 1

Google: What's Good in Your PC Is Good in Your Pocket

Few companies in the world enjoy the overwhelming name recognition that Google does. And why shouldn't it? Many people consider Google's Web search to be the gateway to the Internet itself, a portal to absolutely anything they're looking to find. Google applications (Gmail, YouTube, Google Talk, Documents — the list goes on) are often essential tools for business and pleasure alike. Indeed, chances are good that you and people you know have their browser's home page set to www.google.com.

For many people, e-mail and the Web are essential elements in their daily lives. And now we can add a new addiction to another marvel of technology — a marvel whose sophistication and dominance now rival that of the Internet itself — the cellphone. This diminutive device travels everywhere to keep people connected to friends, family, and (unfortunately) the office. In some cases the cellphone has taken the place of the old-fashioned wired telephone line. Modern cellphones rival the computing power of full-fledged desktop computers from just a decade ago, and some data services make it possible to use the Internet on a cellphone at speeds that give at-home broadband connections a run for their money. The possibilities are endless.

So what is Android? (Drumroll, please.) It's a Linux-based smartphone operating system that can run on all sorts of phones. It just so happens to be a particularly cool operating system with some interesting features and an incredible cross-section of industry and community support. At its core, however, Android is simply an operating system.

More importantly, Android is the operating system that powers the T-Mobile G1 — in fact, the G1 is the very first Android-powered phone to be sold anywhere in the world! For you, that means your wireless companion benefits from everything Android has to offer. Hang on — it's going to be a fun ride.

Unveiling Android the Google Way

For some time, Google has recognized the power of the cellphone. In fact, did you know that Google has tailored versions of its home page for different devices? It also makes mobile Gmail, Maps, and other programs available for free to users of a variety of phone models. Any iPhone owner can tell you how helpful Google Maps is in your pocket or purse — especially when you're trying to track down the address of that killer Chinese restaurant across town.

But, like any good company, Google isn't content to rest on its laurels. Making a Web page that fits on your cellphone's screen is a worthy task, but Google knew that it could do more. In 2005, it purchased Android, a Silicon Valley start-up company that had been quietly and secretively working on creating software for the next generation of cellphones. The company continued working in stealth mode under Mother Google's watchful eye until November 2007, when Android was officially unveiled to the world.

Google makes Android available for free, like most of its products. At first glance, this doesn't seem like a healthy way to turn a profit, but Google has a plan. The company recognizes the importance of cellphones (and more generally, anything that fits in a pocket and can connect to the Internet) in its business strategy in the coming years; with Android, the company has its own platform for deploying Google-branded services without having to deal with other vendors.

Google looks at Android as yet another way to suck you into the bountiful Google ecosystem of products and keep you there — and for anyone using Gmail or Maps on a daily basis, that's not a bad thing at all.

What Android Is (and Is Not)

Just like your desktop or laptop computer, your cellphone — no matter how big or small — runs an operating system. The operating system is the brain of your phone: the software that "talks" to the phone's processor and other hardware, manages memory (you have to put those ringtones *somewhere*, right?), and allows applications (such as mobile Web and e-mail apps, music players, and games) to do their thing.

This tiny operating system varies significantly from phone to phone and from manufacturer to manufacturer. Cellphones broadly fall into one of two categories based on the kind of operating system they use: plain ol' phones, which are sometimes playfully called *dumbphones,* and *smartphones,* which can be expanded. The line between the two is blurry and can change slightly depending on who you ask, but the distinction is important nonetheless.

The dumbphone

Regular phones are devices built from the ground up with a certain set of capabilities in mind, and the options for expansion beyond that are limited. In a way, you can think of their operating system as a walled fortress with no entrance or exit; the manufacturer (sometimes in cahoots with your wireless carrier) decided what software would be allowed to run when the phone was created, and that's that. You may have limited capability to add games or small applications, but these items are usually limited in what they can do because the operating system restricts them.

You might be saying "Down with dumbphones, then — let's move on to the good stuff!" Hold on for just a second, though; it's not all doom and gloom. Most phones sold today, such as the Motorola RAZR and LG Chocolate series, are in this category, and just because their capability for expansion is limited doesn't mean they're not already capable devices. Here are some dumbphone advantages:

- ✔ **They're connected:** Many modern phones have support for e-mail; have full Web browsers; and can feed you with weather, news, sports scores, and more.

- ✔ **They keep you entertained:** Frequently, these phones offer music players, games, powerful cameras, and video playback.

- ✔ **They just work:** Because you can't install just any old application you want on these devices, they tend to be more stable and less buggy than smartphones. The manufacturer and network carriers can test every conceivable configuration of the software before it's sold to you and me.

See, dumbphones aren't so bad!

The smartphone

A *smartphone* is simply a cellphone with a standard operating system and a capability to create and install new programs. It frequently features advanced input systems, such as QWERTY keypads or large touch screens, and impressive features such as integrated GPS for mapping your location and Wi-Fi networking for speedy Web browsing and e-mail access. And, just like the

PDAs of yore, smartphones with the same operating system are often available from multiple manufacturers, making it easier to select the hardware that meets your needs and preferences.

At the end of the day, dumbphones are limited in what they can and cannot do, and that's what ultimately drove the creation of the smartphone. You can think of the smartphone as the spiritual successor to the personal digital assistant (PDA) — the Apple Newtons and Palm Pilots of the 1990s — combining a PDA's capabilities with a phone into a single, pocketable bundle of convenience.

This list describes a couple of benefits that made those PDAs so useful:

- ✔ **They were "open" for development.** Hobbyists and giant corporations alike could create their own software for the devices and then distribute those applications to the world. If you wanted a better note-taking program, for example, odds are someone had already created one that you could buy or, in some cases, download for free.

- ✔ **The experiences were consistent across devices.** If you didn't like Palm's hardware, you could go buy a Sony Clié — but you didn't have to relearn everything about using it because it still used the Palm OS. Similarly, you could switch between a Casio Cassiopeia, an HP Jornada, and a Compaq iPAQ with aplomb because they all ran Microsoft's Pocket PC platform.

It wasn't all fun and games, though. Early smartphones, such as Handspring's Treo 180 and 270, were compromises. They were neither great PDAs nor stellar phones, frequently forcing users to purchase separate, dedicated, simpler dumbphones for those times when PDA functionality wasn't necessary to have around.

Over time, though, manufacturers have expertly and seamlessly integrated the two devices into a single experience. Battery life is less of an issue than it ever has been (although, to be honest, there's no such thing as *too much* battery life). The inclusion of sophisticated word processing, spreadsheet, and presentation applications have made it possible to take short business trips without hauling the laptop along. In fact, smartphones — once designed for and used almost exclusively by businesspeople — have become so easy and fun to use that they're now frequently marketed to and used by college students, stay-at-home parents, and everyone in between.

And then came Android and the G1

Historically, the still-young smartphone operating system market has been dominated by heavyweights Windows Mobile (owned by Microsoft), Symbian (supported by a consortium of phone manufacturers), and

BlackBerry OS (owned by RIM), with the Apple iPhone's special flavor of Mac OS X more recently shaking the boat. Smartphones are really nothing more than pocket-sized computers, and to a certain extent, the software reflects this: Windows Mobile is a scaled-down version of Windows, and the iPhone runs a lightweight version of the Mac's operating system.

Hmm — Windows and Mac. I sense a pattern here. Sure enough, the same fight for operating system dominance in the PC industry is being waged on a smaller scale on your cellphone. Linux is in the thick of things, too, although no one has managed to create a popular, widely used smartphone operating system that's based on it — and that's where the Android-powered G1 comes in.

Recognizing What Makes Android Unique

With established players such as Windows Mobile already running on all sorts of great, capable phones in the marketplace, fair questions to ask are, "What makes Android any different, and why would you choose it?" Or if you work in an organization where the G1 was provided to you, you might be wondering why *they* chose it?

First, I want to get the "X factor" out of the way — that subjective preference for one product's look, feel, and reputation over another's that has separated Mac users from PC users, Ford buyers from Chevy buyers, and Spartans from Athenians for thousands of years. Similarly, some die-hard BlackBerry users cannot be convinced that an Android phone such as the G1 could ever do the job, and hey, that's just fine. Different strokes for different folks, I always say.

Here are some of the unique features that set Android apart:

✔ **Android is tightly integrated with Google products.** One great thing Android has going for it is its extremely (and I do mean *extremely*) tight integration with much of what Google has to offer. For true Google junkies, this integration could be a deciding factor. Sure, Google offers many of its services to other devices, but none can claim to ooze Google from every nook and cranny the way an Android device does. As an example, Google Maps offers an almost desktop-like experience on Android, right down to Street View support. And the G1 offers one advantage that Google Maps on your desktop never can: You can take it with you.

✔ **Android is thoroughly modern.** It's the only mainstream smartphone operating system out there now with that new-software smell. It may be an unfair advantage, but as the new kid on the block, Android inherently gets to lay claim to the title. In practical terms, this doesn't mean terribly much, but you can expect Android to be best equipped to take advantage of today's most advanced mobile hardware — like the G1, for instance.

✔ **Android enjoys massive support from the developer community.** In other words, if you are looking to add a particular application to your phone, odds are very good that the app already exists, is under development, or at the very least is on a developer's mind somewhere in the world.

The list doesn't stop there, though. Like Windows Mobile, Android enjoys a terrific level of device independence, meaning that you can expect to see it running on all sorts of phones from different manufacturers soon. Believe it or not, that's a great thing for you and your G1: The more people who are using Android, the more companies that will embrace it — and that means more software and more support for everyone.

From an employer's perspective, there's a lot to love about Android, too. Android-powered phones will eventually offer true support for Exchange Servers, the Microsoft-supplied e-mail and scheduling systems used by many of the world's companies. Google also allows manufacturers and carriers to lock down their phones, meaning they can be configured so that no additional software can be added without approval: This restriction is no fun for the user, granted, but it avoids a potential support nightmare for companies looking to deploy hundreds or thousands of units.

Adding the Hardware Component

As huge as Google is, it still can't act unilaterally. No company in the world can create a brand-new smartphone platform and expect phones using it to magically will themselves into existence. Far from it, in fact; it takes a small army of big names to give an initiative such as Android a fighting chance in a world dominated by Microsoft and Apple — and a small army is exactly what Google has cobbled together.

The Open Handset Alliance

If Android itself is the king showing on the table, the Open Handset Alliance (OHA) might be the ace up Google's sleeve. Concurrently with Android's announcement in 2007, Google revealed that it had already secretly convinced tens upon tens of the world's largest manufacturers, network operators, and software companies to sign up and support it. Together, this consortium makes up the OHA, whose primary function is to evangelize Android as a platform and steer its future direction and development.

Some important companies are involved with the Open Handset Alliance, which bodes well for Android's future. This list gives you a closer look at some of these players and their involvement in the alliance:

- ✔ **Sprint Nextel and T-Mobile:** T-Mobile has been an enthusiastic supporter of Android since day one, as evidenced by the fact that it has brought the first Android device in the world to market. Fellow U.S. carrier Sprint Nextel is also in the OHA, so it's likely that they'll be offering their own Android-powered phones before too long as well. So what about AT&T and Verizon? Both have both expressed interest in Android without fully committing to the OHA, so it's possible they'll be offering devices as well.

- ✔ **NTT DoCoMo:** Though most of us have never used an NTT DoCoMo phone, the carrier is Japan's largest, meaning it oversees one of the world's most advanced mobile phone networks. Its involvement bodes well for the creation of extremely powerful and creatively designed Android handsets, though you may need to be in the Land of the Rising Sun to use them!

- ✔ **China Mobile:** This carrier is China's largest — which, as you might imagine, also makes it the world's largest with close to *400 million* subscribers, greater than the population of the United States. There's no telling what sorts of Android devices China Mobile might launch, but it can mean only good things to have the world's biggest carrier on your side.

- ✔ **HTC:** The HTC name may not be familiar, but it's a huge Windows Mobile licensee, and odds are you've seen, used, or owned one or more devices created with HTC's involvement. If you have a G1 in front of you right now, you definitely have! AT&T Tilt, Sprint Touch Diamond, and T-Mobile Dash and Wing are other examples of devices manufactured by HTC. Historically, HTC has sold only phones running Windows Mobile, so adding Android to its stable is a big deal.

- ✔ **LG, Motorola, and Samsung:** Together, these three phone manufacturers account for more than a third of all cellphones sold globally. That's a lot of talking!

Why the T-Mobile G1 is important

The T-Mobile G1 is important because you're using it, of course, but other forces are at play here, too. As the first retail Android phone to be sold anywhere, the G1 is an important showcase for the technology and a critical way to get people exposed to everything that makes Android great.

The fact that a major carrier like T-Mobile and a huge manufacturer like HTC jumped headfirst into the Android pool to produce the G1 speaks volumes about the industry's level of commitment to what Google has managed to do. There are hundreds of good reasons for their enthusiasm, and I'll be walking you through a good number of them.

Chapter 2

Powering On and Getting Around

*A*s powerful as the G1 is, you'll be delighted to find that applying power to your little beast and navigating its primary functions really couldn't be any easier. If you're already using a cellphone — any cellphone — several buttons, concepts, and functions will already be familiar to you. And all G1-specific features are intuitive.

In this chapter, you find out about the physical features of the G1. Next, you power on your phone for the first time, unless you've cheated and already done so — I can't say I blame you! Then I walk you through making a phone call — this is a phone after all — and handling incoming calls.

Android Anatomy

Let's begin by taking a look at some of the physical features of the T-Mobile G1 in Figures 2-1 and 2-2. Here are the highlights (keep in mind that Android-powered phones that come to market in the future will share many of these same features, so it'll be a snap to make the transition if you change devices):

Figure 2-1:
The
T-Mobile
G1, screen
closed.

Send   Home   Back   End

Trackball

✔ **Touch-sensitive display:** The display is your phone's main way of conveying information to you. Because the display can register the touch of your finger, you'll frequently find it convenient to navigate the phone's screens and menus this way.

The touch screen relies on contact with skin to function, so you won't be able to use a stylus or fingernail (or wear gloves or mittens) to operate it.

✔ **Menu button:** Pushing this button once will bring up the most commonly used commands and features for the application you are currently using. Pushing it again hides them.

Conveniently, the G1 has also added a Menu key to the keyboard, so there's no need to stretch your weary right thumb to reach the Menu button if you happen to be typing and need access to the current application's options!

✔ **Send button:** If you've used another cellphone in the past, you probably know what this one is all about — you use it to answer or place calls. By holding this button down for a moment, you can use it also to dial by voice, a feature I discuss in Chapter 4.

✔ **Home button:** This button gets you to Android's Home screen in a hurry. Windows users might think of this as the Show Desktop command.

Figure 2-2:
The
T-Mobile
G1, screen
open.

TIP

Pushing and holding Home for slightly longer will show you a task switcher that allows you to select from the last six running applications (see Figure 2-3). This is surprisingly handy! It's not uncommon to need to switch between two or three apps at a time — whether it be on your PC or your G1 — and this app switcher allows you to do that without constantly returning to the Home screen.

✔ **Trackball:** Like many modern BlackBerry devices, the G1 uses a track-ball for navigating around the screen and selecting items. The trackball is an alternative to the touch screen itself, which might be more convenient depending on how you're holding the phone.

✔ **Back button:** This button always returns you to where you just were before you arrived at the current screen. If you've opened a menu, pushing Back closes it; if you've opened an application from the Home screen, pushing Back brings you back home.

✔ **End button:** Like the Send button, users of other cellphones will feel right at home with End. If you are currently on a call, pushing End terminates it. If you are not on a call, pushing End puts your phone to sleep, and holding down End gives you additional options. I discuss all of these shortly.

Figure 2-3:
The task
switcher.

✔ **QWERTY keyboard:** Not everything you do on your phone requires lengthy text input, but when you're writing a long e-mail message, sprucing up a report, or jotting down a novella, nothing beats a full QWERTY keyboard like the one on your PC (except a heck of a lot smaller). Simply slide the display to reveal it.

✔ **Volume buttons:** Like many phones, the volume buttons adjust your ringer volume when you are not on a call and the earpiece volume when you are on a call.

✔ **Camera button:** Pushing and holding the camera button momentarily opens the Camera application, which is a nice shortcut for avoiding the trek through the Applications tab when you want to take a picture. I take a close look at the G1's photographic skills in Chapter 11.

Touch-sensitive display

Touching the display is a key part of the Android experience. Google left nothing to chance and also included support for trackballs and other physical controls, but you'll likely find that you touch the display most of the time.

Throughout this book, I refer to touches and swipes. A *touch,* as you might expect, simply means touching an item on the display. A *swipe* occurs when you touch the screen, hold your finger down, and move it to another part of the screen in one fluid motion.

What's the function of a swipe, you ask? It's used to move an object (say, a menu) from one location to another. It sounds awkward, but it's intuitive once you've used it. Say you have a piece of paper on a table; to move it, you'd just press down on it and move your finger. If you wanted to move it faster and farther, you might flick your finger, let go, and watch the paper keep going. Android supports both types of swipes, and you'll be mastering them in no time.

The trackball

As fun, easy, and effective as touching the display is, sometimes using a more traditional way of getting around Android makes sense. Your G1 features a trackball for this purpose; just roll the ball up, down, left, or right with your finger to navigate.

When you're using the trackball, a highlight appears on the screen to indicate what's currently selected. Once you've selected what you want (a menu item, for example), you press the trackball to activate it.

Turning on the G1 for the First Time

The first time you turn on your G1, you'll go through some extra fanfare that you won't need to worry about ever again (unless you intentionally clear the phone's memory or want to change Google accounts). Just follow the quick instructions on the screen. The entire process doesn't take but a minute or two of your time and then you're on your way.

So, what are these extra steps all about? Because Android is so tightly integrated with Google's services, your phone requires you to have a Google account associated with it at all times, and you'll be asked to get this account configured when the phone first turns on.

Keep in mind that your Gmail login *is* your Google account, so if you already use Gmail, you're in the door — just use that login information here. Don't have a Gmail account (or any other Google account)? No biggie — you can create one right from the G1. Just touch the Create icon during the initial setup process to get started.

Requiring a Google account to use the G1 has numerous advantages; for example, Gmail, Calendar, Contacts, and Google Talk are all automatically configured for you as soon as you finish entering your account information. What's more, if you already use Google services on your PC to manage your contacts and calendar, your information will automatically start downloading to your phone; after a few minutes, it'll all be there! Even better, any future changes you make from your PC or your phone will be synchronized without any intervention on your part.

Because your phone is perpetually tied to a Google account, your most important information — your calendar, contacts, and e-mail — are effectively backed up to Google's servers on a continual basis. Rest easy!

Feels Like the First Time: Resetting Your G1 and Deleting Its Memory

If you need to change the Google account associated with your G1 or you're giving it to someone else (the horror!), you'll want to *hard reset* it. This procedure clears its internal memory and makes it "forget" your Google account so that someone can enter a new account. To perform this operation:

1. **Turn off the G1.**

2. **Push and hold down the End button until a pop-up menu appears (see Figure 2-4). Choose Power off, and then touch OK.**

Figure 2-4:
The Phone options menu, accessible by pushing and holding the End button.

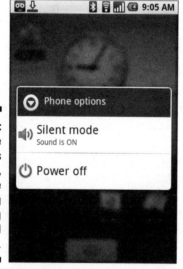

3. **Push and hold both the Home and Back buttons.**

4. **While continuing to hold Home and Back, turn on the phone by pushing and releasing End.**

   Keep holding Home and Back — you're almost there! After a brief self-test (you'll see a lot of pretty colors), a message appears warning you that "this operation will delete all your personal data, and reset all settings to manufacturer default." In other words, all your stuff will be gone from the phone.

5. **Release the Home and Back buttons.**

6. **Push the Send button to go ahead with the hard reset operation, or press anything else to cancel it and power on the G1 normally.**

Although you can always hard reset your G1 and enter in the same Google account you were using before to restore your contacts, calendar, and Gmail data from Google's servers, anything else — applications you've installed, for example — will be gone. Perform this operation only if you're really sure you want to do it.

Changing the Battery

If you're able to charge your G1 every night and you sell (or recycle) the phone within two or three years of purchasing it, you may never need to worry about changing the battery. However, if you find that you're on the phone, browsing the Web, or checking your e-mail for several hours during the day without access to a wall outlet — or the time to use it — you might be well-served to carry a spare battery that can be purchased from T-Mobile or a number of online retailers. When you're in a pinch and the G1 is about to use its last drop of juice, you can just power down, swap in the spare battery, and be on your merry way.

You'll also find that the battery loses its ability to hold a good charge over time — anyone who's used a laptop computer with the same battery for a few years is probably familiar with this phenomenon. If you find that you can't ever get a full day's use out of the phone when you're not using it heavily, you know that it's time to replace the battery.

Changing the battery is pretty simple:

1. **Shut off the phone.**

2. **Open the keyboard.**

   To the left of the keys, you'll notice a crack in the plastic. This is the top of the tab that holds the back cover in place.

3. **With a fingernail, pull outward and down (toward the bottom of the phone) on the crack.**

   Keep pulling! You'll feel a little resistance, but don't worry — you're not breaking anything. If you're doing it correctly, the entire backside of the G1 is coming off as you pull.

4. **Once you've pulled the tab completely free of the phone, pull the back cover out of its tabs on the other end of the phone and set it aside.**

You're now looking as the G1's innards. In the center, the black object labeled *HTC Innovation* is the battery. (If you've already replaced the battery with a third-party model, it could be a different color and labeled differently.)

5. **Using a fingernail, lift up on the battery's tab.**

   The top end of the battery has a small tab with a hole underneath it, which makes it easy to grab hold of the tab with a fingernail.

6. **Pull the battery completely out of the phone and set it aside.**

   At the bottom of the battery chamber, you will see three gold pins — these are the connections that allow the G1 to draw power from the battery.

7. **Take the new battery and place its bottom end (the end with the gold squares on it) inside the G1's battery chamber first, making sure that the gold squares line up with and touch the G1's gold pins.**

8. **Drop the other end of the battery into the chamber and push down until the entire battery is flush with its surroundings — in other words, the battery shouldn't be sticking up.**

   Before you replace the cover, you can power on the phone to make sure that you've installed the battery correctly.

9. **Align the bottom end of the cover (the end without the large tab) so that it's touching the bottom end of the phone. (The tiny tabs on the cover should be in the phone.)**

10. **Rotate the cover into place and press on it until the large tab is back in its original position.**

    As when removing the cover, you'll feel some resistance here — that's okay. Keep pressing until the tab is fully in place.

It *sounds* like a lot of work, but after you've done it a couple times, I promise that swapping the battery will be a cinch — you might even be able to do it standing up. Just don't blame me if you drop anything!

Your First Phone Call

In the following simple example, I show you how to power on the phone and call a friend to talk about your exciting new purchase (or gift, if you're lucky):

1. **Push the End button.**

   The End button doubles as the Power button — on the button, note the small power symbol below the handset symbol. After a moment, the display turns on. Your phone is coming to life! After the display turns on, you see a display similar to the one in Figure 2-5.

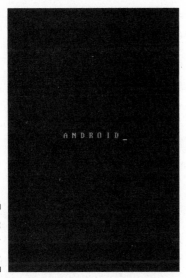

2. Wait for the Home screen to be displayed.

As its name suggests, the Home screen (see Figure 2-6) is your phone's home base, not unlike the desktop on your PC. You see this screen when your phone turns on or when you push the Home button.

3. **Push the Send button.**

No matter which screen you're looking at in your phone, pushing Send opens the Dialer application, which is a fancy way of saying "place your call here."

4. **Touch the Dialer tab at the top of the display.**

You can dial calls in several ways in a phone this sophisticated, but to start, we'll keep it simple by dialing just like our ancestors did, by entering the phone number into a numeric keypad. That's the function of the Dialer tab, shown in Figure 2-7: It's a numeric keypad plus a display to see what you typed. Nothing more, nothing less.

Figure 2-7:
The Dialer application with the Dialer tab selected.

5. **With your finger, use the numeric keypad displayed on the screen to enter the number of your friend (or enemy, if you're looking to make someone jealous).**

6. **Push Send.**

That's it! You should be sharing the news of your newly acquired gem of modern technology as you read this paragraph.

You probably noticed during your call that another screen appeared, giving you additional options and information — this is the call progress screen. Let's take a look at it next.

Call Handling

Android offers all the call handling options that you would expect in a modern cellphone. You can

✔ **Swap calls:** If you have two calls going at the same time, swapping between the two is just a matter of touching an icon.

✔ **Merge calls:** Bring two active calls together into a conference line.

✔ **Add a call:** Make a second phone call while keeping your first call on hold.

✔ **Hold the call:** This option allows you to place a caller on hold, but he or she won't get any relaxing background music while waiting.

✔ **Mute the call:** This is similar to hold, but you can still hear the caller — they just can't hear you.

✔ **Enable the speakerphone:** Did you know your G1 can function as a full-fledged speakerphone? It's great for an impromptu meeting around the lunch table, and it's just a touch away.

Let's say you've still got your friend on the line from the preceding section, and you're still looking at the call progress screen — the place Android takes you during a phone call. (If you've already hung up, go ahead and call back — your friend won't mind!) But now you'd like to add a second buddy into the conversation. With Android, it's a piece of cake:

1. **While in the call progress screen, push the Menu button.**

   A grid of options available for the call appears at the bottom of the display, as shown in Figure 2-8.

2. **Touch Add call.**

   This returns you to the Dialer (you may recognize this screen from the call you placed earlier). As before, we'll keep it simple by entering a number directly into the keypad using the Dialer tab at the top of the screen. And don't worry — we'll cover the rest of those tabs soon enough!

3. **Using the Dialer tab, dial the second number and push the Send button.**

   The call progress screen returns, but this time two boxes are shown. The uppermost box shows you the number you just called and displays "Call in Progress" below it, letting you know that this is the person you're currently talking to. Below that, a second smaller box reminds you that you still have your first call on hold! Let's rectify that.

4. **Push the Menu button (or press the Menu key on the keyboard if that is more convenient).**

Figure 2-8:
Call options.

5. **Touch Merge calls.**

   A new box appears on the screen indicating that you are now in a conference call.

   Now, what if another friend wanders into the room while you're having this lovely conversation? No problem — it's easy to get the third person in on the call, too:

1. **Push the Menu button or press the Menu key on the keyboard.**

2. **Touch Speaker.**

   Note that the gray bar below the Speaker icon turns from gray to green, indicating that the G1's speakerphone is enabled. The screen also holds a second clue: a new icon appears in the status bar to the left of the signal strength indicator (see Figure 2-9).

Figure 2-9:
The
speaker-
phone
is on.

Speakerphone

In Android, the status bar is almost always visible. In case you leave the Dialer application while you're still on a call, the status bar icon will still be there to remind you that everyone can hear you!

When you've finished your chat, just push the End button. Now, repeat the process until every one of your friends knows all about your fabulous G1. When you're ready, I'll be waiting for you in the next chapter.

Chapter 3

The Home Screen and Status Bar

*W*hen it comes to Android, there's no place like home. No, seriously — you'll be spending much of your time at your G1's Home screen. Just like your PC's desktop, the Home screen is a jumping-off point for everything that your device has to offer — but beyond that, it's a full-fledged application unto itself that can keep you informed and entertained with a minimum of fuss.

In this chapter, I discuss in detail what the Home screen is all about, what it can do for you, and how you can customize it to make it your own. Then you take a look at the Home screen's slimmer cousin, the status bar, which you'll find in most of the screens that you encounter.

The Home Screen: Not Just for Checking the Time Anymore

Many people who carry average phones will tell you that their handset's Home screen is good for one thing, and one thing only: checking the time. Yes, that's an important function — especially as wristwatches become an ever rarer accessory — but with the G1, the beautiful clock is just the beginning.

Applications, shortcuts, and widgets

The Home screen holds three kinds of items:

- ✔ **Applications:** These are exactly as they sound. You can take any application on your phone and make an icon for it right on your Home screen, giving you quick and easy access to your most frequently used programs.

- ✔ **Shortcuts:** Think of shortcuts as files on your PC. Shortcuts are not applications but links to data within them. For example, you can create a Contact shortcut that links directly to a particular contact in the Contacts application or a Bookmark shortcut that opens a specified link in the Browser.

- ✔ **Widgets:** These might be the most interesting thing going on in the Home screen. Widgets, such as the Google search widget depicted in Figure 3-1, are tiny applications that live right on the Home screen and provide information or offer a bit of functionality without having to open a full-fledged app. These are equivalent to gadgets in Windows Vista or — you guessed it — widgets in Mac OS X.

Figure 3-1:
The Google
search
widget.

The Home screen can accommodate all three kinds of items simultaneously, so don't be afraid to experiment! A well-conceived layout that matches your personality and lifestyle can improve your productivity by shaving minutes of needless fiddling with your phone.

The default layout

Out of the box, Android's default Home screen layout, shown in Figure 3-2, is useful without any customization.

The default layout provides the following:

- ✔ **The status bar:** You'll find this little buddy on most of the screens in Android, giving you tidbits of information that you'll always want access to. I dive into the status bar's capabilities later in the chapter.

✔ **A large analog clock:** This widget reminds me of the clocks in my elementary school classroom. It's legible, looks great, and does a good job of making sure that I'm not tardy. The digital version is available just above in the status bar.

✔ **myFaves, Dialer, Contacts, Browser, and Maps icons:** If you're signed up to a T-Mobile plan that includes myFaves, you'll see an icon for it here — it appears automatically! Other than that, Google assumes that Dialer, Contacts, Browser, and Google Maps are the four applications you'll be spending most of your time using (and it's probably an accurate guess), so it drops icons for them front and center on your Home screen.

✔ **The Applications tab:** This tab is the portal to your device's library of applications. Unless you have an icon for a particular app on your Home screen, you'll access the application from here.

Figure 3-2:
The Home screen's default layout.

Getting around the house

To select something on your Home screen, simply touch it or select it by pressing the trackball while it is highlighted. Different types of Home screen items react differently when you select them:

✔ Applications and shortcuts take you somewhere. For example, touching or selecting the Browser icon opens the Browser.

✔ Widgets may allow you to interact with them, they may take you somewhere, or they may do nothing at all. I discuss Widgets in a moment.

You can do more than select things you see on your Home screen, though. If you don't see the application you want to open, expanding the Applications tab as in Figure 3-3 will reveal it (along with every other application installed on your phone).

Figure 3-3:
The Applications tab expanded.

You can expand the Applications tab one of two ways: either touch and hold the tab and then physically drag it across the screen, or simply touch it briefly. To close the tab once it's expanded, touch it again, or drag it in the opposite direction that you opened it.

By swiping the Home screen left or right, you reveal two additional but empty Home screens. They're there simply to give you more room to add applications, shortcuts, and widgets that may not be important enough for the first page. These secondary pages are always just a swipe away.

The only thing that makes the primary Home screen "primary" is that it's between the two secondary screens. That way, you can access either of your secondary screens when you are looking at your primary screen simply by swiping. From anywhere in Android, one push of the Home button takes you back to the last thing you were looking at on your Home screen, and two pushes of Home return you directly to your *primary* Home screen.

Making your house a Home

The default layout's great, but who's to say that Dialer, Browser, Contacts, and Maps are the four icons you want front and center? Let's make a few changes to your Home screen.

Change the wallpaper

First, let's swap out that tired wallpaper for something a little more engaging:

1. **Push the Menu button.**

   The Home screen's menu appears, as shown in Figure 3-4. Keep in mind you can always press the Menu key on the keyboard to accomplish the same task.

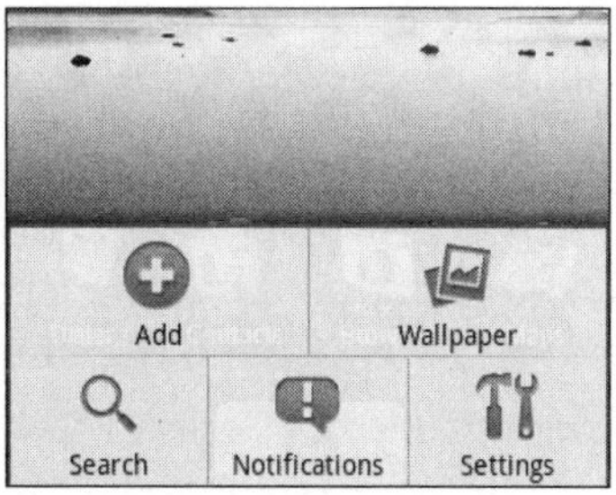

Figure 3-4:
The Home
screen
menu.

2. **Press Wallpaper.**

   You're presented with a menu of wallpaper sources.

3. **Press Wallpaper gallery.**

   You see a viewer that allows you to quickly navigate and preview all the available wallpapers on your device. In this view, you can swipe the picture strip at the bottom to move it quickly to the right or left. The highlighted picture — which always appears in the middle of the strip — is shown magnified at the top of the display.

4. **When you've settled on your new wallpaper, press Set wallpaper.**

Remove an application icon

Looking better already! Next, remove the Dialer icon, since you can get to the Dialer application just as easily by pushing the Send button:

1. **Touch and hold the Dialer icon.**

   After a moment, the icon becomes larger and the phone briefly vibrates, indicating that the icon is ready to be moved to another location on the screen or deleted. The normal arrow icon on the Applications tab changes to a trash can icon, telling you that this is where you drag items you want to remove.

2. **Without lifting your finger, drag the icon to the trash can icon and release.**

The Dialer icon disappears, but don't worry; the application itself is still there. You can access it by pushing Send or, like any application, by choosing its icon in the expanded Applications tab. (You can also add a Dialer icon back to your Home screen at any time by holding down on the icon from the Applications tab and dragging it back out to the Home screen.) Deleting or moving around an application's Home screen icon — similar to an alias in Mac OS or a shortcut in Windows — has no effect on the underlying program.

Move a widget

That clock might look a little better in the center of the screen, don't you think? (Maybe not, but for the purposes of this exercise, just politely agree — it's easy to change back, I promise!)

1. **Touch and hold the clock.**

   As with the Dialer icon, the clock becomes larger, indicating that it can be moved or deleted. You'll also feel a quick, gentle vibration.

2. **Drag the clock until you're satisfied with its location.**

3. **Release your finger.**

The clock stays where you left it, and because it won't move without being first held down, it's difficult to accidentally jostle it out of place.

Add an application icon

Next, let's assume that you find yourself eating out a lot, so you simply *must* have quick access to a calculator to compute tips. Easy:

1. **Push the Menu button.**

2. **Touch Add.**

   The different categories of items you can add to the Home screen appear. Besides the wallpaper you've previously selected, you can choose from applications, shortcuts, and widgets.

3. **Touch Application.**

4. **Touch Calculator.**

Now you have a Calculator icon right on your Home screen — perfect for all that number crunching you intend to do.

You can add an application icon in another way, and this method gives you more control over the icon's location:

1. **Touch the Applications tab to expand it.**

2. **In the Applications tab, touch and hold the Calculator icon.**

3. **Drag away from the icon's original location, making sure that your finger stays in contact with the screen.**

A copy of the icon stays with your finger while the Applications tab closes.

4. **Release your finger on the Home screen where you'd like the icon to appear.**

Add a widget

Finally, let's add a Google search widget (this is a Google Android phone, after all). There's a problem, though — your primary Home screen's getting a little crowded! In Figure 3-5, you see what happens if we try to drop the widget on the primary screen.

Figure 3-5: Whoops, no more room!

This is exactly where those extra Home screens come in handy:

1. **While on the primary Home screen, swipe left.**

   You should now be looking at a blank (or nearly blank) Home screen, devoid of the icons and other goodies you're used to. Notice that you can still see your wallpaper in the background.

2. **Push the Menu button.**

3. **Touch Add.**

4. **Touch Widget.**

   The menu expands to show you all the widgets that are available to you.

5. **Touch Search.**

   And the Google search widget magically appears on the screen, ready for action.

But wait — we have one small problem. If you swipe to the right (so you're now looking at the primary Home screen) and then swipe right again, you'll notice that T-Mobile has already done you the favor of putting the same widget here! No problem; let's delete it. The process is identical to removing an application icon:

1. **Touch and hold the widget.**

   The widget gets bigger and the phone briefly vibrates, indicating that the widget (just like an icon) can be moved or deleted.

2. **Keeping your finger held to the screen, drag the widget to the trash can icon and release.**

Your masterpiece is now complete. In addition to a modified primary Home screen, you can now kick off a Google search (see Figure 3-6) without even opening the Browser first!

Figure 3-6:
Searching
from the
Home
screen.

When it comes to Home screen organization, everything is treated the same. You can move, delete, and add applications, shortcuts, and widgets with equal aplomb using the guidelines you've uncovered here.

The Status Bar: Not Just for Checking Signal Strength Anymore

By now, I hope you agree that Android has taken the Home screen to an exciting new level of configurability and utility. But the fun doesn't stop there! If there's one element of Android you'll see more than the Home screen, it's the status bar — and Google has put plenty of work into making sure it's the best darned status bar it can be.

Most phones have a bar across the top of the phone's screen that shows basic, critical information about the phone's functions. Signal and battery strength are the usual suspects, while many handsets also show indicators for new text messages and voice mail messages.

Android's status bar does all those things and much, much more, thanks to its *notification system,* which I discuss in the next section. First, though, look at how your G1's status bar is laid out. Figure 3-7 shows a typical status bar.

Figure 3-7:
The Android status bar.

Elements of the status bar include the following:

- ✔ **Notification icons:** These icons appear when various applications on your phone are trying to get your attention. I discuss these shortly — they're powerful and important and you'll use them frequently.

- ✔ **Silent mode:** If you've set your G1 to keep quiet, you see a musical note with a slash through it.

- ✔ **Data transfer:** When you're connected to T-Mobile's (or another wireless carrier's) data network, arrows indicate when data is flowing in and out of your G1. When that network is a high-speed 3G network, you also see a small *3G* symbol with these arrows. (But trust me, you'll know from the speed of your e-mail and browsing that you're on a fast connection!) 3G connections aren't available in all areas, so if you see an

E here instead, you'll know you're connected to EDGE, an older, slower network. If you have Wi-Fi enabled and you're currently connected to an available network, the arrows are replaced with the symbol shown in Figure 3-8.

Figure 3-8:
The G1 displays a special symbol in the status bar when you're connected to a Wi-Fi network.

WiFi active

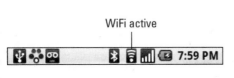

- ✔ **Signal strength:** A standard signal strength meter indicates how well you're connected to the cellular network. The more lit bars, the better.

- ✔ **Battery meter:** This icon indicates how much battery life remains on your current charge — green means full; black means it's time to find an outlet! A lightning bolt in the icon means that your device is currently being charged.

- ✔ **Time:** Because you can never have too many ways to find the time, Android includes a digital clock in the status bar in addition to the clock widget available for the Home screen. Because the status bar is visible on almost any screen, the time is always available to you.

Notifications

You can think of *notifications* as brief messages that applications on your phone send to you, usually because they have something important to tell you. For example, the Messaging application might need to let you know that you've received a new picture message from Mom, or the Dialer app might need to tell you that you still have an active call.

The nature of the notification varies from application to application, and as you install new applications, you'll sometimes find that they add their own notifications into the mix. Fortunately, the status bar is adept at aggregating and organizing these notifications into a one-stop shop.

Perhaps the best part of these notifications is that because they're displayed in the status bar — which you see in almost every application — you receive them no matter where you are in your G1. Whether you're checking e-mail, browsing the Web, or buying music through Amazon MP3, you'll get that text message that the meeting has been moved up an hour. (You'd better get going!)

Notification icons

The first way to display a notification, and the least invasive, is through a simple icon that appears in the upper left of the status bar. What if you need to see more, though? What if a simple icon isn't enough to convey what you need to know? For that, you can expand the status bar to show a complete list of current notifications, as in Figure 3-9.

Figure 3-9: The expanded status bar, showing notifications.

As you can see, you had dinner scheduled with a friend at 5:00 p.m. sharp today — and thanks to your G1, you weren't late for it! Fortunately, the Calendar always makes sure that you're kept abreast of upcoming appointments by posting notifications to you.

Beyond icons: The expanded view

To get to the expanded view of the status bar, hold your finger over the status bar and pull downward, as though the bar were a rolled-up window shade. You can do this from any screen in Android in which the status bar is visible (which is most of them).

To open an application that has posted a notification, just tap the notification in question. The behavior of the application will vary based on the type of notification and the design of the application itself. However, in general, you'll be taken to a screen that provides more detail regarding the notification and gives you an opportunity to act on it.

To close the expanded view, just hold your finger on the bar at the bottom of the screen and drag it upwards. Once the bar reaches approximately the midpoint of the screen, you can let go and it will roll back up until you need it again.

Chapter 4

Making and Receiving Calls and Text Messages

*S*ure, the G1 can do lots of amazing things — but at the end of the day, it's still your phone, too! And as phones go, the G1 is about as full-featured as they come, offering conference calling, voice dialing, tight integration with your contacts, and a whole lot more.

These days, though, much of the world's real-time communication doesn't involve a single spoken word. SMS, or Short Message Service, lets virtually any two cellphones zip quick text messages to each other, which is great for those times when a full-on phone conversation isn't needed or you're in a quiet (or noisy) place where talking to a caller isn't appropriate. SMS is massively popular — so popular, in fact, that it has given rise to its own shorthand language designed to better accommodate the 160-character limit imposed on each message. (Let's just say that "c u l8r" isn't a license plate number.)

Android has comprehensive support for text messaging using both SMS and its more advanced sibling, MMS, or Multimedia Message Service. As its name implies, Multimedia Message Service allows you to attach pictures, sounds, and videos to your messages, and it also does away with that pesky character limit that SMS messages have to deal with.

In this chapter, I show you how to make the most of your phone calls and your text messages alike. I may not turn you into an SMS lingo wizard (I recommend `http://www.laptopmag.com/advice/tips/become-a-text-message-shorthand-pro.aspx` for that), but at least you'll know where and how to send that lingo!

The Dialer

In Android, pretty much everything you do takes place in an application — just like on your PC — and making and receiving a phone call is no different. You use the Dialer application (pretty obvious name, right?) to take care of that business on the G1.

I touch on the Dialer in Chapter 2, where I show you how to make a phone call using the numeric keypad and introduce you to some of its cool call handling features. Now let's break down the application and find out everything it can do.

First, we need to open the Dialer. As with any other application, you can do this by tapping its icon in the Applications tab (or tapping its icon on your Home screen, if you've kept it there). The thing is, Google (and everyone else in the world) considers dialing an important function for a phone, so you can access the Dialer also by pushing the G1's Send button.

Once you're in the Dialer, you're presented with four tabs across the top of the screen:

- **Dialer:** This is as simple as it gets — all you have is a numeric keypad (just like on any other phone you've used) and a bar where the entered number is displayed.

- **Call log:** If you enter the Dialer application by pushing the Send button, this tab is selected by default. It shows you a list of recent calls, both made and received. The Call log tab also shows you calls that you've missed.

- **Contacts:** This is your address book, the place where you store and access contact information permanently.

- **Favorites:** Think of this as the Contacts tab for people you *really* like (or those you have to call a lot). It's basically just a special subgroup of the Contacts list for people you need to get in touch with frequently.

The Dialer tab

As simple as the Dialer tab seems, it still has some tricks up its sleeve. While entering a number, the backspace icon to the right of the number bar (the area where the number appears) allows you to erase mistyped digits. Holding down the icon for about a half second, though, will erase the entire number.

If you need to make edits in the middle of a number, there's even a way to do that: Use the trackball to highlight the number bar, and then move the trackball left or right to position the cursor. Then backspace to delete digits, or touch number keys to add new ones.

When you're ready to place the call, either touch the number bar or push the Send button.

What if you're provided a phone number as a series of letters, like 1-800-ANDROID? (Don't dial that, by the way; I just made it up.) You can always play hunt-and-peck like you would with a traditional phone, looking for each letter below the digits on the screen. The G1 makes it easier, though:

1. Open the screen.

The screen flips from portrait (tall) to landscape (wide) mode, the numeric keypad disappears from the screen, and a prompt appears telling you to "Use keyboard to dial," as shown in Figure 4-1.

Number bar   Backspace

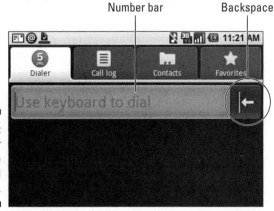

Figure 4-1:
The Dialer tab with the keyboard open.

2. Enter the full phone number, typing letters exactly as they appear.

For example, if the number were 1-800-ANDROID, you would actually type **1800ANDROID**. There's no need to translate the letters to their corresponding numbers — Android does it for you!

Although the backspace icon is still visible on the screen when the keyboard is open, the backspace key on the keyboard performs exactly the same function, so you may want to use that instead. (No need to move your weary thumbs all the way from the keyboard to the screen if you don't have to, right?)

Checking voice mail

As with many cellphones, the G1 lets you know that you've received voice mail by displaying an icon of a cassette tape at the top of the screen. In Android's case, the voice mail icon on the status bar is a *notification*, which means you can get additional information about it by holding your finger on the status bar and pulling downward (see Figure 4-2).

Figure 4-2:
The New voicemail notification.

The easiest way to check your voice mail is to simply pull down on the status bar and touch the row that displays New voicemail, which immediately dials your voice mail number. You can dial voice mail also directly from the Dialer tab by touching and holding the 1 key on the screen for about a half second. This is particularly useful if you don't have any voice mail waiting but want to change your greeting or other options.

Jotting down a number and turning it into a contact

It happens all the time: You're meeting someone for the first time, you need the person's phone number, and he or she rattles it off to you. But what if you don't have pen and paper at the ready? With Android, that's no problem at all — you can quickly enter the number immediately using the Dialer without going through the process of creating a new contact first.

1. **From the Dialer tab, enter a phone number as though you're going to dial it.**

2. **Push the Menu button or press the Menu key on the keyboard.**

   You see a single menu item at the bottom of the screen — Add to contacts.

3. **Touch Add to contacts.**

   A new screen appears asking you whether you'd like to add the number you just typed to an existing contact or create a new contact altogether.

I go through the creation and management of contacts in detail in Chapter 8. For now, keep in mind that you can add numbers to your contacts in this way.

The Call Log tab

Android considers the call log so important, it's the first thing you see if you open the Dialer application using the Send button (see Figure 4-3). It makes sense — we all miss calls from time to time, receive calls from numbers we don't already have stored in our contacts, and return calls to people who've recently called us. The call log helps you accomplish these tasks quickly and easily.

Missed call

Received call

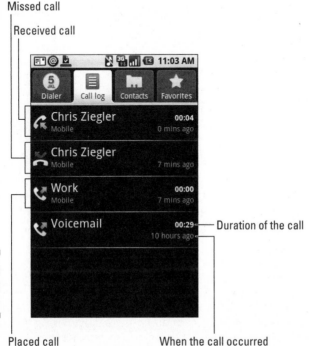

Duration of the call

Figure 4-3:
The Call Log tab.

Placed call   When the call occurred

Let's take a look at the components that make up the call log. For phone numbers that appear in your contacts, the name associated with the contact appears for each line item along with the type of number (mobile, work, home, and so on); otherwise, you simply see the phone number. To the left of the name or number, you see one of three icons:

- ✔ **Placed call:** A green arrow moving away from the phone. These are calls that you made.

- ✔ **Received call:** A blue arrow moving toward the phone. These are calls made to your phone that you (or whoever was using your phone) answered.

- ✔ **Missed call:** A red arrow bouncing off the phone. If you missed a call (or rejected it by pushing the End button), you see this.

To the right of the name or number, you see how long ago the event occurred. For made and received calls — in other words, calls that you didn't miss or reject — you also see the duration of the call in hours, minutes, and seconds.

You can call back any number in your call log by simply touching it, but the fun doesn't stop there. Touch and hold any item in the log for a moment and you're presented with a menu of options — things you can do with that item. Let's have a look at those options:

- ✔ **Call <number>:** This is the simplest option in the menu; it immediately dials the number as your G1 received it. This option is the same as touching an item directly in the call log.

- ✔ **View contact:** If the selected name is already in your contacts, this menu item becomes available and sends you to a screen with the details for that contact. This is especially handy if you want to call someone back at a different phone number than the one on which they called you.

- ✔ **Edit number before call:** If you need to make some changes to the number before you can dial it — adding a country code, for example — choose this option. The number will be sent to the Dialer tab for your editing.

- ✔ **Send SMS message:** This option opens the Messaging application and sets you up to send a text message to the selected phone number. I cover text messaging later in this chapter, in the "Text Messaging Application" section.

- ✔ **Add to contacts:** If the selected number isn't currently in your contacts, this menu item appears. It takes you to a screen that allows you to add the number to an existing contact or a new one (just like touching the Add new contact option in the Dialer tab).

✔ **Remove from call log:** Selecting this menu item removes the selected number from the log. How covert of you!

Pushing the Menu button from within the call log reveals a single item, Clear call log. It immediately clears your entire log without asking you for confirmation. Use caution!

The Contacts tab

I'll get into all the great things you can do with the Contacts tab in Chapter 8, but for the purposes of making phone calls and sending text messages, there are a few things to note here.

When you open the Contacts tab you see a list names, organized alphabetically, as shown in Figure 4-4.

Figure 4-4:
The Contacts tab.

To call a contact, touch the contract's row in the list. You are presented with a list of all numbers associated with that contact. In fact, you'll see every number twice — once under the Dial number heading and once under the Send SMS/MMS heading. This allows you to quickly place a call or send a text message with a single tap.

Touching a number under the Dial number heading places the call immediately. If it's more convenient for you to keep your thumb on the trackball, you can also roll it until the number you want to call is selected, and then either press down on the trackball or push the Send button.

Similarly, selecting a number (either with the trackball or by touching it on the screen) under the Send SMS/MMS heading opens the Messaging application and sets up a new message ready to be sent to the selected number.

If you return to the Contacts tab — remember that you can do this by pushing the Back button — there is another way to place a call or send a message to a contact without entering the contact's screen. Touch and hold the contact's row in the list and you'll see a menu of things you can do with the contact, just as you saw in the call log. I look at most of these items in Chapter 9, but notice a couple of entries here:

- **Call <number type>:** The contact's default phone number appears here. For example, if Joe's default number is his cellphone, this says Call Mobile. Touching it places a call to that number immediately.

- **Send SMS/MMS:** Touching this starts the Messaging application and gets you set up to send a text message to the contact's default number.

The Favorites tab

The Favorites tab functions in the same way as the Contacts tab — except it's different. Bear with me, though! Of all the people in your contacts list, how many do you call on a regular basis? Five, maybe ten percent? Of course, that ratio varies from user to user, but the point is that you don't want to have to sift through a thousand contacts every time you want to find Mom's number. (What, you don't know your own mother's phone number? Shame!) The Favorites tab simply gives you a secondary view of your contacts, boiled down to just the ones you handpick.

Turning a contact into a favorite

Adding a contact to your Favorites is easy:

1. **Touch the Contacts tab to open it.**

2. **Touch and hold on a contact that you want to turn into a favorite.**

   After a moment, a menu appears.

3. **Touch Add to favorites.**

You can also touch a contact (or hold down on a contact and select View contact from the pop-up menu) to go into its details screen and touch the star in the upper-right corner — it changes from white to yellow, indicating that it is now a favorite (see Figure 4-5).

Going to the Favorites tab, you'll see that the contact has a yellow star next to it. I recommend not adding any more contacts here than you have to — the idea is to make this list short and easy to browse; otherwise, the Favorites tab becomes no faster to use than the Contacts tab itself.

Figure 4-5:
The high-lighted yellow star means this contact is a favorite.

Removing a favorite

Removing a favorite is similar to adding one:

1. **Open the Favorites tab.**

   You can also open the Contacts tab if you prefer — favorites are listed in both tabs.

2. **Touch and hold the contact whose favorite status you want to remove.**

   After a moment, a menu appears.

3. **Touch Remove from favorites.**

You may also touch the star in the upper-right corner of the contact's details screen (accessible either by touching the contact or by touching and holding the contact and choosing View contact from the pop-up menu), turning the star from yellow to white.

Removing a favorite doesn't delete the contact — it just removes it from the Favorites tab. The contact is still accessible from the Contacts tab.

Controlling the Volume

The volume buttons on the side of the phone control different types of volume, depending on what you're doing on your G1 at the moment. For example, during a call, the volume buttons adjust the volume of the phone's earpiece, speakerphone, or your headset — whatever you happen to be using. If you're listening to music, the buttons control the volume of the music.

If you're not actively performing any task that involves the use of sound, the volume buttons control the volume of the ringer that is sounded when you receive a call or a message. When you first press one of the volume buttons, a window appears that displays a horizontal bar graph indicating the ringer's loudness (or softness). Touching up or down on the volume controls increases or decreases the ringer volume, respectively, and changes how "filled up" the bar graph appears. You also hear a sample tone with each press to give you an idea of how loud your G1 is set to ring.

Do you want the phone to vibrate or do nothing when a call comes in? Repeatedly push the minus volume button until the phone reaches zero volume (that is, the bar graph is empty); at that point, the phone is set to briefly vibrate when you get a call. If you push the minus volume button one more time, the phone won't vibrate *or* ring; it'll be completely silent when a call comes in. Indicators showing a phone set to a low volume, to vibrate, and to silent mode are shown in Figure 4-6.

Figure 4-6:
Ringer, vibrate, and silent modes, respec- tively.

Bossing Your Phone around with the Voice Dialer

Sometimes, fiddling around with the screen to dial a number or to find a contact isn't convenient. Maybe you're walking briskly down the street and are fearful (understandably) of bumping into a huge, scary dude walking the other way if you take the time to look down and set up a call. There's got to be a better way, right? There is!

The G1 supports voice dialing through a special application called Voice Dialer, which can understand spoken commands to dial a phone number, dial a contact by name, dial voicemail, or redial the last number dialed. What's more, Voice Dialer can hear you from a distance of a few inches, so if you intend to make a call on the G1's speakerphone and you want to use the Voice Dialer, you don't need to begin by holding the phone up to your ear.

To command the G1 by voice, you need to get it to listen. That's easy:

1. **From any screen or application, push and hold the Send button.**

   After a moment, you hear a beep, and the Voice Dialer window shown in Figure 4-7 appears. As the message on the screen indicates, you can speak any of a number of different commands:

   - Call <contact name>

   - Call <contact name> at <phone number type>, such as "Call Chris Ziegler at work."

   - Call voicemail

   - Dial <phone number>

   - Dial an emergency service such as 911, customer service using 811 or 611, or information using 411

   - Redial

2. **Speak a command, with the handset held up to your ear or in your hand. Speak in a clear, deliberate voice.**

   You see a list of possible matches for what you've said (see Figure 4-8). This list always appears, even if the G1 thinks you could have said only one thing, as a final safeguard against dialing someone you didn't intend to!

3. **Touch the selection that matches what you want to dial, or touch Cancel to end the command without doing anything.**

Figure 4-7:
The Voice
Dialer.

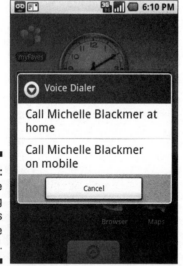

Figure 4-8:
Possible
matching
commands
in the Voice
Dialer.

Using Wired and Bluetooth Headsets

The G1 includes a wired stereo headset with a microphone attached to it that you can plug into the ExtUSB port on the bottom of the phone (see Figure 4-9). A special version of the Mini-USB standard for connecting devices between one another, ExtUSB was developed by the G1's maker, HTC, to support both data and audio connections. What does that mean to you?

Basically, you use this port to connect the G1 both to your computer and your headset — there's no separate audio jack like you may find on some other phones.

Figure 4-9:
The G1's
ExtUSB
port with
protective
rubber cover
removed.

Wired headsets

To use the wired headset, simply connect it to the ExtUSB port. The G1 will automatically use the headset as your hands-free device — any calls you make or receive will be routed through it until you disconnect it.

Pairing and using Bluetooth headsets

For true portability and convenience, nothing beats a wireless Bluetooth headset for making hands-free calls. In this section I walk you through the process of setting up a Bluetooth headset for use with the G1.

The G1 doesn't support the use of stereo Bluetooth headsets for listening to music wirelessly, but it does fully support mono headsets for making phone calls. Some wireless stereo headsets can function in a mono mode, however, so some of these may work with the G1. Consult your headset's manual, or just give it a try — there's no harm in that!

As with any other phone, the first step in using your Bluetooth headset with the G1 is to pair it. To do this:

1. **Set your headset to pairing mode.**

   The procedure for doing this varies from headset to headset, so check your manual.

2. **On your G1, go to the Applications tab and select the Settings application.**

3. **In the Settings application, touch the Wireless controls menu item.**

4. **In the Wireless controls screen, make sure that the third item down — Bluetooth — has a check mark, as shown in Figure 4-10.**

   If the option doesn't have a check mark, add one by touching the check box. If you've just added a check mark, it'll take a moment for the G1's Bluetooth radio to power on, at which point you'll see a new icon for Bluetooth in the status bar (it looks like an angular letter *B*).

Figure 4-10:
The
Wireless
controls
screen of
the Settings
application.

5. **Touch the fourth menu item in the Wireless controls screen, Bluetooth settings.**

   The G1 begins scanning for Bluetooth-capable devices around you with which it can pair and then shows them in the lower half of the screen under the Bluetooth devices heading.

6. **Touch the item in the Bluetooth devices list that represents the headset you're trying to pair.**

   Generally, the name you see in the list will be the make or model or both of your headset. You will be able to identify headsets in this list (versus Bluetooth-equipped computers and other devices) also from a headset icon, which appears to the right of the device's name.

7. **Using the keyboard, enter your headset's Bluetooth passcode in the field that appears on the screen.**

Generally, the passcode for headsets is 0000 or 1234, but check the user manual for your specific headset.

8. **Touch OK.**

You're all set! Connected appears below your headset's name in the Bluetooth devices list, and the Bluetooth icon in the status bar changes, as shown in Figure 4-11, to indicate that a device is connected.

Bluetooth device name   Headset icon

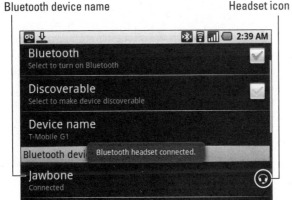

Figure 4-11: A Bluetooth headset has been connected.

At this point, your phone operates just as though a wired headset were connected — all call sounds are routed through your Bluetooth headset as long as it's turned on. If you turn the headset off, the G1 recognizes that and switches back to using its own earpiece and microphone (in other words, you'll be holding the phone up to your face to make or receive calls). When you turn the headset back on, the G1 recognizes that, too, and starts using it immediately (as long as the G1's Bluetooth radio is enabled, as mentioned in Step 4 in the preceding list).

Unpairing Bluetooth headsets

Now that you have your Bluetooth headset paired to the G1, the two will automatically hook up whenever they're both turned on and the G1's Bluetooth functionality is enabled. But what if that's not what you want? If you have another phone that you want to use with this headset instead, you may want to unpair it from the G1 so that it won't connect to the headset anymore. Fortunately, that's a cinch:

1. **Go to the Applications tab and select the Settings application.**

2. **In the Settings application, touch the Wireless controls menu item.**

3. **Touch Bluetooth settings.**

   In the lower half of the screen below the Bluetooth settings heading, you'll see your paired headset — even if the headset is turned off or is too far away from your G1 to connect. The G1 remembers the headset because it was paired previously.

4. **Touch and hold the headset's menu item until a pop-up menu appears.**

5. **Touch Unpair.**

The Messaging Application

As I mention in the previous section, you can automatically start a text message to a contact or a phone number in the call log in several ways. If you open the Messaging application from a contact's entry or an entry in the call log, you are taken to the thread for that contact (see Figure 4-12). A *thread* is simply all the text messages sent between you and the other person, organized into a timeline — a conversation view. This stellar feature, which many cellphones don't have, makes text messaging a lot more fun and useful by organizing text the same way that you're used to seeing instant messages using Google Talk, AIM, or a similar application on the PC (that is, in a back-and-forth conversational style).

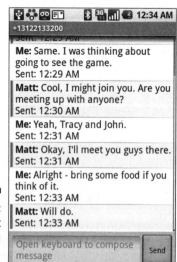

Figure 4-12:
A text
messaging
thread.

Sending and receiving text (SMS) messages

For our first example, let's assume you've selected a contact or call log entry from the Dialer application, which brings you straight to the thread view for that contact in the Messaging application. You don't have to specify who this message is going to — that information has already been filled out for you.

Composing a message

If you have the screen closed, you'll notice at the bottom of the application that you need to "Open keyboard to compose message." In other words, you need to use the keyboard to type. Otherwise, all you can do here is read the conversation; you can't send anything new.

Sending a text message from a thread couldn't be any simpler:

1. **Open the screen.**

   The display switches to landscape mode and the text at the bottom changes to "Type to compose, press Enter to send."

2. **Start typing.**

3. **When you've finished composing your message, press the Enter key on the keyboard or touch Send on the screen.**

   While the message is in the process of sending, an envelope icon appears to the right of your message in the thread. When your message has been successfully sent, the envelope disappears.

When you receive a text message, you receive a notification that looks like a postcard in the status bar. The message also briefly flashes across the status bar, which lets you preview it without having to go into the Messaging application.

Pull down the status bar to reveal the notifications screen, which gives you details on who sent the message and a snippet of its contents. Just touch the notification in the screen to hop over to the Messaging application and view the thread.

Remember how I said that SMS messages must contend with a 160-character limit? The G1 actually allows you to send messages that are longer than that limit — they're just broken into multiple 160-character chunks. Android gives you a way to see how many chunks you're sending, too. As you get close to breaking past the first 160 characters, a small gray box appears in the upper-right corner of your text box showing two numbers separated by a slash. The first number is the number of characters you have remaining in the current chunk; the second is the total number of chunks that will be sent. Because some phones (especially those without thread views) make it tricky to view multiple chunks, this is a handy piece of information to have.

Starting a new thread from the Messaging app

Of course, you don't need to use the Dialer application to start a new message; you can do it directly from the Messaging application as well. Let's give it a shot:

1. **From the Applications tab on the Home screen, touch the Messaging icon.**

   Many people consider text messaging to be a vital part of their communication; if you're a member of that group, you may want to place the Messaging icon on your Home screen as well.

2. **No matter what screen within the Messaging application you are brought to, push the Menu button and choose Compose.**

   When you open the Messaging app, you are always brought back to the last screen you had open, whether it be the list of all threads or a specific thread. Regardless, the Compose menu item is always available.

   At this point, you should now be looking at an empty thread with a blank To box at the top.

3. **Open the keyboard.**

4. **Touch the To box and start typing the name of a contact or a phone number. Separate multiple contacts with commas.**

   As you type, the phone searches for possible matches in your contacts list. If it finds some possibilities, it displays them in a list below the To box. At this point, you can select the intended recipient from the list, or if you're typing a phone number, you can continue to type the entire number. If you've selected a contact from the list, a comma is automatically placed at the end and you can immediately begin typing additional contacts if you so choose. If you're typing a number, however, you have to enter the comma yourself.

Now, the process for sending a message is no different than if you had selected the contact from the Dialer app — just type your message using the box at the bottom of the screen and press Enter or touch Send when you're finished.

Touching and holding for about a half second on an individual message in a thread calls up a pop-up menu of available commands. Depending on the message, here are some of the items you may see:

- ✔ **Call <number>:** The phone number that sent the message to you is displayed here, and you're given the option of calling it back. Yes, with actual voice instead of a text message!

- ✔ **Forward:** This option effectively forwards the selected message by inserting it into a new thread and allowing you to choose a new recipient.

- ✔ **View message details:** This item displays the type of the message (SMS or MMS), the sending phone number, and the date and time the message was sent.

- ✔ **Delete message:** If you just want to delete a single message out of the thread, this is the way to do it.

Sending and receiving multimedia (MMS) messages

Sometimes, text isn't enough — you want to send a picture of that celebrity who just walked by you in the restaurant. (He doesn't look so great in person, does he?) Maybe you want to send a clip of a song that's on the radio right now. With MMS, the possibilities are almost endless.

In Android, MMS messages and SMS messages are treated virtually the same, which makes them especially easy to send and receive. Both types of messages appear together in your threads — the only difference is that an MMS message can have more stuff attached to it. Figure 4-13 shows an example of an audio file attached to an MMS.

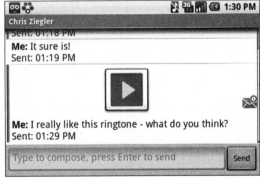

Figure 4-13: An MMS message with an attached audio file.

Multimedia messages can be composed of as many as three parts, all of which are optional:

✔ **Subject:** This is a short subject line that describes your message.

✔ **Body:** The body is the main text of your message. This can be rather long, unlike the 160-character limit for SMS messages.

✔ **Attachment:** This can be audio, video, or a picture.

Sounds kind of like an e-mail message, doesn't it? Think of an MMS message as a way to push a short, quick e-mail to someone's phone so that he or she will see it as quickly as possible — your recipient is notified right away.

Unlike Android-powered cellphones like the G1, not all cellphones support MMS (though most phones sold today do). You may want to consult with your friends and other contacts before sending multimedia messages to them. If you send a message that a recipient claims never to have received, his or her phone may not support MMS.

Playing media attached to a multimedia message

What happens when you get a multimedia message? Not much, actually — you receive a notification that looks like a postcard, just as you would with a text message, and you view the message in the same way. The difference, of course, is that there's likely some multimedia involved! When you open a message thread after receiving an MMS, you might see a picture or a play button, depending on the type of media that was sent to you. If you see a picture, touching it enlarges it; if you see a play button, pushing it plays the clip. Waiting a moment or pushing the Back button returns you to the thread.

Sending a multimedia message

When sending a multimedia message, you enter text just as you would with a text message — this becomes the body of the message. Additionally, you can enter a subject and attach media.

While you're looking at a message thread (create a new one if you like), push the Menu button. Of the menu items, two are notable here:

✔ **Add subject:** As mentioned, subject lines are one of the features that set MMS messages apart from SMS messages. Touching this item converts your message to MMS format and gives you a subject line at the top of the screen — just use your keyboard to fill it in.

✔ **Attach:** This is the fun part! Touching this displays a prompt to select a type of media to attach to your message.

If you touch Attach, several attachment types are available in the menu:

✔ **Pictures:** This item opens a viewer that allows you to select a picture that is already stored on your G1.

- ✔ **Camera:** Like Pictures, this option ultimately results in a picture being attached to your message. However, this option opens the Camera application and allows you to take a picture right now instead of selecting an image that you already have stored.

- ✔ **Audio:** Here, you are prompted to select an audio file that resides on your phone. Highlighting a file without selecting it plays it, allowing you to preview the clip before you decide whether to attach it.

- ✔ **Record audio:** This option is the audio equivalent of Camera — it allows you to record a new audio clip in real time instead of selecting a clip you already have stored.

- ✔ **Slideshow:** Want to send several pictures at once? Check out the Slideshow option! It lets you assemble multiple pictures into an automatically advancing slideshow — basically a movie of pictures.

After you've selected an attachment, you see a view similar to Figure 4-14, which shows an attached audio clip that's ready to send. Using the buttons to the right of the attachment, you can play the clip, replace it with a different one, or remove it. When you're ready to send your message, do so just as you would with a text message: Touch Send on the screen or press the Enter key on your keyboard.

Figure 4-14:
An audio clip attached to a multimedia message, ready to send.

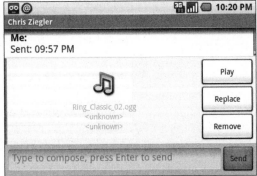

Managing threads

From any thread, you can push the Menu button and choose All threads to return to the view showing all your message threads (see Figure 4-15). Remember that a thread is a messaging conversation that you're having with a single person (or a group of people), so the All threads view has one line item for each contact (or group) with whom you've been sending or receiving messages. A green bar appears to the left of any thread that has messages you haven't read.

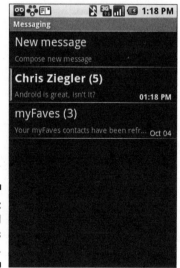

Figure 4-15:
The All
threads
view.

From here, you can start a new thread with the New message row at the top of the screen, or you can view a specific thread by touching it. You might open a specific thread to review what has been sent and received within a conversation or if you want to send a new message to a contact with whom you already have a thread going.

To delete an individual thread:

 1. **Touch and hold a thread's row in the list for about a half second.**

    A new menu appears with two options, View and Delete. Choosing View has the same effect as simply touching a thread's row — it opens the thread for viewing and sending new messages.

 2. **Touch Delete.**

    You are warned that the entire thread will be deleted. Gone. Kaput!

 3. **Touch OK.**

To delete all threads from the All threads view:

 1. **Push the Menu button.**

 2. **Touch the Delete threads menu item on the screen.**

    Just like when you're deleting a specific thread, Android warns you of the implications of your action. But in this case, you're about to wipe all text and multimedia messages from the phone. Be sure you want to do this!

 3. **Touch OK.**

Messaging settings

Pushing the Menu button from the All threads view reveals a Settings menu item, where you can tweak some of the finer points of how the Messaging application works. Let's have a look at those settings. First, you have a section titled SMS settings with a pair of options:

✔ **Delivery reports:** Selecting this option (by touching the check box to add a check mark) instructs T-Mobile that you want to receive confirmation that every message you send has been received. To view this report, touch and hold an individual message in a thread, and choose View report in the menu that subsequently pops up.

✔ **Manage SIM card messages:** The Subscriber Identity Module (SIM) card in your phone — the little chip that lets T-Mobile know who you are — has a bit of memory on it and is capable of storing SMS messages. This feature is rarely used, but if you have messages on your SIM that you'd like to view or move to your G1, this is the place to do it.

Next, you have a section entitled MMS settings:

✔ **Delivery reports:** This option works just like the Delivery reports setting for SMS messages.

✔ **Read reports:** In addition to getting a report when an MMS message is delivered, you can be notified when it has been read by the recipient.

The Read reports option doesn't always work — it depends on the recipient's phone and the whims of the wireless carriers involved. Don't count on it!

✔ **Auto-retrieve:** Unlike an SMS message, MMS messages don't necessarily have to be automatically retrieved in their entirety. Downloading them can consume considerable amounts of data because of the media attachments involved, so if you're trying to conserve battery life or watch your data usage, it's a good idea to turn off this option. (Don't worry — you'll still be able to retrieve your MMS messages manually.)

✔ **Roaming auto-retrieve:** When you're using a carrier other than T-Mobile — in a foreign country, for example — data services can be *extremely* expensive. It's a good idea to leave this option disabled so that your MMS messages aren't quietly retrieved at exorbitant rates when you're roaming on another carrier's network.

Taking your G1 abroad

You may be familiar with the concept of *roaming*, which occurs when you use your cellphone on a carrier's network other than the one to which you subscribe. Roaming occurs whenever you take your G1 to another country. (Although T-Mobile USA is a subsidiary of the same company that operates T-Mobile networks in several other countries around the world, each of these networks is considered its own independent entity.)

The G1 supports quad-band GSM, which means it supports four frequencies of cellular service that together cover most of the globe. It's a handy feature, but be careful — roaming expenses can add up *very* quickly! Using your phone to make voice calls outside the United States can cost from 49¢ to as much as $5 per minute or more, so roaming is best left to the expense account if possible.

Data services such as MMS cost even more — which is why the G1 gives you the option of disabling the automatic retrieval of such messages when the phone is not connected to your own network. This option can definitely save you from experiencing a heart-stopping expensive bill or two. Within the United States, you may roam on carriers besides T-Mobile, but don't worry — domestic roaming through T-Mobile doesn't cost you anything.

The final section is called Notification settings:

- ✔ **Notifications:** Deselecting this option stops you from receiving a notification in your G1's status bar when you receive a message. Typically, you'll want to leave this enabled, unless you get a ton of messages and it's a distraction.

- ✔ **Select ringtone:** Choosing this item brings up a menu that lets you choose the tone your G1 plays when you receive a message. (Regardless of your choice, you won't hear it if your phone is in Silent mode, which is a good thing!)

- ✔ **Vibrate:** If this option is selected, the phone rumbles when you receive a message.

Now that you have the two important ways to communicate with the G1 under your belt, I turn my attention to searching the Internet, your contacts, and your applications in Chapter 5 — but communication is never far away! I revisit it with Gmail in Chapter 6 and instant messaging in Chapter 12.

Part II

Putting the "Google" in Google Android

The 5th Wave   By Rich Tennant

"Well, here's what happened—I forgot to put it on my 'To Do' list."

In this part . . .

You can't look too far through the G1's software and hardware without finding at least a few references to Google — and for good reason. Like all Android-powered phones, the G1 gives you access to a world of Google-based services that you're accustomed to using on your desktop or laptop computer. Part II takes you through these, starting with searching (where else?) and moving on through Gmail, Maps, Contacts, and Calendar.

Chapter 5

Searching

*I*n the past, searching for stuff on your phone — be it your sister's phone number, your dry cleaner's address, or the name of the 23rd president of the United States (it's Benjamin Harrison, by the way) — wasn't just difficult. Often the task was simply impossible, depending on the model of phone you were using.

Imagine, for a moment, having to use your PC without any searching capability whatsoever. What would you do? How would you get around? How would you find everything you needed on a daily basis? It'd be nearly impossible considering the sheer volume of information we're presented with these days.

Cellphones are becoming so powerful and capable that they're starting to run into some of the same challenges we've faced on PCs. No longer can a self-respecting phone get by without powerful contact and rich Internet searching capabilities. The information's there — why shouldn't we be able to find it easily?

If you guessed that a phone powered by software created by the world's largest Internet search company would have fantastic searching capabilities, you're absolutely correct. The G1 can pretty much search its way out of even the tightest bind, and in this chapter, I show you how.

Searching the Internet

If your previous phone had a small screen, a slow data connection, or both, you may never have even thought of searching the Internet from the comfort of your mobile companion. With the G1, though, it's a way of life.

Searching from the Home screen

You have no fewer than three good ways to search the Internet right from your Home screen. Yes, *three* — and if that's not a sign that Google's trying to tell you it's fun and easy to find what you're looking for out on the Web, I don't know what is!

All of the techniques mentioned in this chapter (and indeed, any technique mentioned in this book where text entry is needed) require that the keyboard be open so that you can type your search. So you may as well prepare by opening the screen now if you haven't already.

Method one:

1. **Push the phone's Menu button or the keyboard's Menu key.**

   The Home screen's menu appears.

2. **Press the Search icon.**

   If you have a Google Search widget somewhere on your Home screen (your primary screen or elsewhere), you are taken straight to the widget, and a keyboard cursor is placed in the text box so you can start typing. If you don't have the widget, a search box appears at the top of the screen below the status bar, as shown in Figure 5-1.

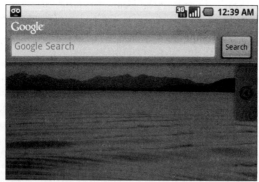

Figure 5-1: The Google search bar.

Method two:

1. **Press the search key (see Figure 5-2) on the keyboard.**

   That's it. There is no Step 2. That search key's great, isn't it?

Figure 5-2:
The key-
board's
dedicated
search key.

Search

If you have the Google Search widget installed somewhere on your home screen, you can kick off a search from the Home screen with a third method:

1. **Locate the widget.**

   Remember that by default, the Google Search widget is installed on the portion of the Home screen to the right of the primary view, so you'd get to the widget by swiping to the left.

2. **Touch the text box within the widget.**

   This action highlights the text box and places a cursor within it. You're ready to rock.

Searching from the Browser

I cover the G1's amazing browsing capabilities in detail in Chapter 10, but I thought I'd mention it here as well because you search from the Browser in the same way as you do from the Home screen. You can kick off a search from the Browser by pushing the Menu button and choosing Search, or by pressing the search key on the keyboard. Either way, a search bar appears at the top of the application and you can begin typing.

A cool shortcut for initiating an Internet search works only in the Browser, and only when a text field is not highlighted: Just start typing. The application automatically brings up the search bar and shows what you've typed so far. (The text you're typing here can also be a Web site address instead of a search — the Browser detects the difference.)

Starting the search

After you've entered your search terms (don't be shy — enter the same terms you would on your desktop browser), press the Enter key on the keyboard or touch the Search icon next to the text box to put Google to work. (Note that the Search icon is replaced with a Go icon if you've used the Browser shortcut I mention in the preceding Tip.) The Browser opens and you see search results just like you see when you use `google.com` on any computer.

Search suggestions

Regardless of where or how you initiate your search across the vast repository of information we call the Internet, Google provides suggestions of what it thinks you might be trying to search for as you type. (The shortcut I mention in the Browser is a notable exception to this rule.) Although the G1 has an excellent keyboard for its size, it's always great when the phone can help you cut back on a few keystrokes!

For example, if I type *andr* and pause for a moment, the G1 talks to Google's servers and discovers that I might be trying to spell out *android, andrea bocelli,* or *andrew sullivan* (see Figure 5-3). I can keep scrolling down this list to see other suggestions as well, but in this case, it has correctly guessed that I wanted to search for Android. All I need to do is select it from the list and the search will begin immediately.

Figure 5-3:
Search
suggestions.

Searching Your Contacts

The G1 can store a virtually limitless number of contacts, and if you've been using your Google account extensively, you already have a ton of 'em stored on the phone after the very first time you powered it on, signed in, and allowed it to sync up. Sure, you *could* browse through them every time you need to choose one for a call, a text message, or an e-mail, but thankfully you don't have to.

The easiest, most hassle-free way to search your contacts doesn't even involve going into the Contacts application — you do it right from the Home screen:

1. **Start typing the first few letters of the first or last name of the contact you're looking for.**

   You are transported to the Contacts application, and the letters you've typed appear at the bottom of the screen. (In Figure 5-4, for example, I typed *b*.) The contacts list looks and behaves the same way as it always does; the only difference is that it is now filtered to match your criteria.

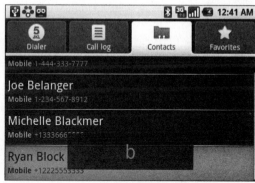

Figure 5-4: Typing a letter of a contact name to filter the contacts list.

2. **Select the contact you are looking for as you normally would.**

   At this point, you can perform all the functions you would expect with a contact, such as placing a call by pushing the Send button or editing the contact.

You can search your contacts in the same way by first going into the Contacts application (or going into the Dialer application and selecting the Contacts tab) and then typing your search. Searching this way requires that one extra step of going into the Contacts app, so you probably won't use this technique unless you're already in the Contacts application when you need to search.

Searching for Your Applications

Fresh out of the box, the G1 starts its life with plenty of applications. By the time you load it with a hearty helping of third-party software, the contents of the Applications tab can become downright daunting to navigate, as shown in Figure 5-5.

Figure 5-5:
A fully loaded Applications tab with a number of third-party apps installed.

How can you manage it? It helps that the list of icons is sorted alphabetically by default, but it probably comes as no surprise that you can search it as well:

1. **From the Home screen, touch the Applications tab to open it.**

2. **Type the first few letters of the application you're looking for.**

   In effect, this search works the same as searching for a contact. As you type, the list gets shorter and shorter. Even if you know only the first letter of the app, that should be enough to narrow the list enough so you can easily see what you're looking for.

With their Google pedigree, it should come as no surprise that Android and the G1 offer such a powerful set of searching tools — but that's just the beginning! In upcoming chapters, I show you how to take full advantage of your contacts, your Internet browsing capabilities, and all the applications you've installed (and can install). The sky's the limit, and we're just getting started.

Chapter 6

Using Gmail and Other E-Mail Services

*L*ike it or not, e-mail is quickly becoming a predominant form of communication in the world. (Actually, let's face it — it's already there.) A healthy percentage of the phone calls that were made or the letters that were mailed ten years ago have now been replaced by e-mails, so we need easy access to our e-mail accounts in as many places (and in as many ways) as possible.

Although many phones can be set up to send and receive e-mail, few can do it with the finesse of the G1. Even if you're coming from an e-mail powerhouse such as a BlackBerry or an iPhone, you'll find that the G1 can do some cool tricks that you may not be used to. Here's a quick look at some of the things that makes Android's e-mail support so great:

✔ **Threaded conversations:** If you use Gmail on your PC, you're familiar with this feature (though you may not know it by name). In a *threaded conversation,* all e-mails from the same conversation are grouped into a single chain, which makes the conversation much easier to read — no more jumping from message to message.

✔ **Push support:** The G1's Gmail application supports *push,* which simply means that new e-mails are fed to you in near real-time instead of you manually retrieving them (or setting them up to be retrieved at a regular interval). In practical terms, it means your e-mail is delivered to your G1 as quickly as possible.

✔ **Notifications:** Most phones can notify you of new e-mail in one way or another, Android has a fabulous notification system integrated into the status bar — and as you might expect, the Email and Gmail applications support it.

✔ **Support for labels and stars:** Just like Gmail on your desktop, you can see what labels are applied to your e-mails and apply or remove labels. You can do the same with stars, which are simply a tool to make important e-mails stand out.

Although your G1 can support any POP3- or IMAP-based e-mail account, note that you have to use a different, dedicated e-mail application for all non-Gmail accounts, which I describe in the second half of this chapter. Gmail gets its own special application (it's a Google service, after all!) that provides quite a bit of extra functionality — including the threaded conversations, push, and label and star support features just mentioned — so use a Gmail account as your primary e-mail account on the G1 if possible.

In this chapter, I introduce you to both the Gmail and Email applications and take you through the full range of their capabilities. Heck, by the end of this chapter, you may be asking yourself whether you even need a PC for this stuff. (Not really, but almost!)

Gmail

Before you even start using Gmail on the G1, it already has one great thing going for it over other e-mail services: There's no setup! Remember when you first turned on the phone and entered (or created) your Google account information? Gmail is an integral part of your account, and it was automatically configured at that very moment.

Retrieving new e-mail

Because Gmail on the G1 uses push technology, there's nothing to retrieving new e-mails — literally. When you choose the Gmail icon from the Applications tab to fire it up, you are immediately taken to the Inbox, where your e-mail is waiting for you (see Figure 6-1).

Sender

Subject

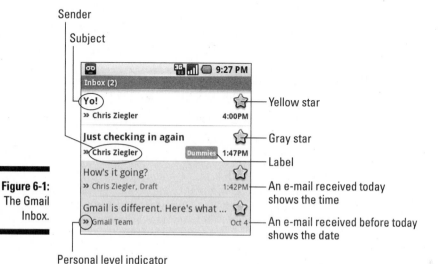

Yellow star

Gray star

Label

An e-mail received today shows the time

An e-mail received before today shows the date

Figure 6-1: The Gmail Inbox.

Personal level indicator

Let's break down what you're seeing here:

✔ **Current label:** If you've used other e-mail services or an e-mail program for your PC such as Outlook, you might be accustomed to organizing your e-mail in folders. (And if you aren't, I'd wager you have a pretty messy inbox!) In Gmail, labels take the place of folders. In many ways, labels function the same way as folders — you can select a label name to see all the e-mails that have that label applied, for example. But you can apply multiple labels to an e-mail. For example, if you get an e-mail about the annual company picnic, you can apply both Work and Fun labels.

At the top of the screen you see the label whose contents (that is, e-mails) you're browsing. The number of unread e-mails with this label applied appears in parentheses after the label name.

In Gmail, even the Inbox is considered a special type of label. In fact, if you load the contents of another label, and an e-mail that's currently in your Inbox appears there, you'll see an Inbox tag in the lower right of the e-mail's row — just as you would for any other label.

✔ **Subject:** The subject line of the e-mail.

✔ **Personal level indicator:** If you have this feature enabled in Gmail on your desktop, you'll be familiar with how it works. A double arrow means the e-mail is addressed just to you, a single arrow means you appear in the To list with other recipients, and no arrow means you don't appear in the recipient list at all (for example, if you received the e-mail because you're on a mailing list or you were blind carbon copied).

✔ **Sender:** The name of the e-mail's sender. If multiple e-mails are in this thread, you will see multiple names here (separated by commas) followed by the total number of emails in the thread in parentheses. If you're in the middle of a reply to this thread, you also see the word *Draft* listed as one of the senders; this simply means your reply has been drafted but not yet sent.

✔ **Label:** Any labels applied to this e-mail, separated by commas. Labels appear in a colored box so they're especially easy to see. Note that you will not see a label's name here if you are currently in that label's view. For example, if you're looking at the contents of a label titled Work, you will not see the Work label applied to any e-mails. Google does this to reduce screen clutter.

✔ **Star:** A yellow star indicates that this e-mail is starred; a gray star indicates that the e-mail is not. Like the Inbox, the star is a special label, which means a view in the Gmail application allows you to see only starred e-mails. How (and whether) you decide to star e-mails is up to you.

✔ **Date or time:** If an e-mail was received today, you see the time it was received. Otherwise, you see the date it was received instead.

Notice that unread e-mails appear in bold with a white background, whereas read e-mails appear in a standard typeface with a gray background.

Although your e-mail is being refreshed in near real-time, there may be times when you're feeling *really* impatient (we've all been there). If that happens, you can refresh your e-mail list manually:

1. **From the Inbox view, push the Menu button.**

2. **Touch the Refresh icon.**

   In the status bar a circular arrow briefly appears to let you know that the phone is synchronizing your information with Google. After the icon disappears, new e-mail received since the last synchronization (no more than a few minutes) appears, the read/unread status on all your e-mails is updated, and labels and stars are refreshed.

Even when you're not in the Gmail application, new e-mails are received and you are notified about them, as shown in Figure 6-2. Just touch the New email notification to go straight to your Inbox.

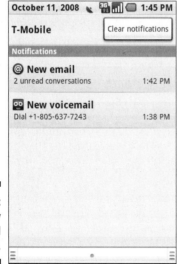

Figure 6-2:
A New
email
notification.

Reading and working with e-mail

The simplest and likely most frequent thing you'll want to do with e-mail is read it. To do this, highlight an e-mail in the list with the trackball and select it by pressing down on the trackball, or simply touch the e-mail with your finger.

Once you've selected an e-mail, the view of an e-mail thread is very much like what you see in desktop Gmail — a series of individual e-mails in succession that can be expanded or collapsed (see Figure 6-3). When an e-mail is collapsed, all you'll see of it is a gray header that tells you who sent the e-mail, whether it's starred, and the date (or time) it was sent. At the top of the screen, you see the subject line, which is common to all the e-mails in the thread.

To read through the thread, swipe up and down or move the trackball up and down. When expanded, the bottom of each e-mail in the thread has three icons — Reply, Reply all, and Forward — which work just as you're used to.

Even though Gmail intelligently shows you all the e-mails pertaining to the same subject in a single thread, it does make a difference which Reply, Reply all, or Forward icon you touch! For example, if you have three e-mails in a thread and you expand the middle message and touch Reply, the text quoted below the body of your reply comes from that middle message, *not* the most recent message in the thread.

Figure 6-3:
A thread
with two
collapsed
e-mails
(left); a
thread
with one
expanded
and one
collapsed
e-mail
(right).

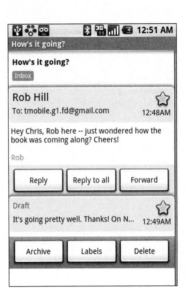

If you scroll to the bottom of a thread that's in your Inbox, you'll see three additional icons: Archive, Labels, and Delete. These work the same as the Archive, Change labels, and Delete menu items, which we look at next. If you're in another labeled view besides the Inbox, you may see different or fewer icons, depending on what options are appropriate. For example, the Archive icon may be replaced with a Remove label icon that simply removes the current label. If you're viewing the Trash, Delete obviously doesn't make much sense, so you won't see it.

Now, return to the Inbox. (If you're currently looking at an e-mail thread, just push the Back button to return to the Inbox.) By touching and holding on a thread's row in the list, you are presented with a menu of things you can do with the thread, as shown in Figure 6-4.

Let's have a look at the available options:

- **Read:** The first option takes you to the thread view to read the thread. This option is equivalent to simply touching the row itself, so don't waste your time with this one!

- **Archive:** This option removes the entire selected thread to the All Mail view, which means the thread is no longer visible from your Inbox but is still in the background if you ever need it. If you're a current user of Gmail, you're probably familiar with this technique.

- **Mark read/Mark unread:** If you've already read a thread, this option lets you mark it as unread, and vice versa. As mentioned, unread messages are in bold type with a white background and read messages are in regular type with a gray background.

✔ **Add star/Remove star:** This option is self explanatory, and like the Read menu item, there's not much point in using it from here — you can simply touch the star itself to the right side of each thread's row.

✔ **Delete:** This menu option moves the entire thread to the Trash. In Gmail, the Trash is deleted on a rolling 30-day cycle, so you have basically a full month to retrieve deleted stuff if you have a change of heart.

✔ **Change labels:** This option lets you modify the labels that are applied to this thread. I get into label management in more detail next.

✔ **Report spam:** If you've received unwanted, unsolicited bulk e-mail (affectionately known as *spam*), touch here to let Google know that the message falls into this category.

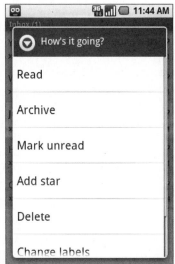

Figure 6-4:
The e-mail
pop-up
menu.

 Staying on top of spam reporting is important because it helps Google do a better job of automatically filtering the garbage before it ever gets to you. That means fewer spam e-mails will show up in your Inbox — and when you're trying to manage your e-mail from the palm of your hand, you don't need any unnecessary distractions!

 All of these menu items are available also from within e-mail threads themselves — just push the Menu button to call up the in-thread menu. To see the Report spam option, you need to touch the More item.

Managing labels

You can't create or delete labels from the Gmail application on your G1; you need to log on to Gmail at gmail.com from your PC or the G1's Browser for that. What you can do, though, is add and remove the labels you've already created. There are two ways to do this, depending on what you're currently looking at. If you're viewing a list of e-mails (say, your Inbox or another label):

1. **Touch and hold an e-mail until the pop-up menu appears.**

2. **Touch Change labels.**

If you're currently within a thread, the procedure is slightly different:

1. **Push the Menu button.**

2. **Touch the Change labels item in the menu that appears at the bottom of the screen.**

Either way, you see a screen that looks similar to Figure 6-5.

Figure 6-5:
The Choose labels window.

Labels that are currently applied to this thread appear with a red circle with a minus symbol within it, and labels that are not applied have a green circle with a plus symbol. As you may have guessed, touching a red circle removes that label, and touching a green one adds it.

Remember that Inbox is just a special label, which is why it appears here. Removing the Inbox label will cause the thread to disappear from your Inbox, but it will still be visible from the All Mail view or by viewing any other labels that may be applied to the thread.

When you've finished adjusting the labels for this thread, touch OK to confirm or Cancel to discard your changes.

Views

The main reason to label e-mails is to tame your wild Inbox into a bunch of more manageable views. All labels in your Gmail account — both built-in labels such as Inbox and Starred and labels you create on your own — have their own views that allow you to see *only* e-mails with that label. The G1 makes it easy to see and use these views:

1. **Go to your Inbox (or any other view).**

2. **Push the Menu button.**

3. **Touch the View labels menu item.**

    You'll see a screen similar to Figure 6-6.

Figure 6-6: The View labels screen.

4. **Touch the name of the label (or other view) whose contents you want to see.**

   In addition to the Inbox, you'll see other views in addition to the labels you've created — Sent, Outbox, Drafts, All Mail, Spam, Trash, and Chats (an archive of your Google Talk conversations, which I explore in Chapter 12).

Views with unread e-mails appear with a vertical green bar to the left and a gray box to the right, containing the number of unread e-mails.

That's all there is to it — you now know how to view any label's contents. You'll notice that every view looks pretty much the same as the Inbox, and you manage e-mails in these views in the same way as you do in the Inbox.

After you've finished working with label views, you'll probably want to return to your Inbox since it's the home base of your vast e-mail empire. To do so, push the Menu button and touch the Back to Inbox item. Or you can touch the View labels item to return to the list of all labels (which includes the Inbox).

Writing and sending e-mails

Of course, there's a whole lot more to e-mail than just *reading* it, isn't there? You'll want to send a few, too, especially when you take the G1's lovely, expansive keyboard into consideration.

If you're replying to (or forwarding) an e-mail you've received, simply touch Reply, Reply all, or Forward to get the process started. If you want to compose a fresh message from scratch, however, that's easy too:

1. **While looking at the Inbox or another label view, push the Menu button.**

2. **Touch the Compose menu item.**

At this point — whether you're replying to, forwarding, or starting a new e-mail, you'll be looking at a screen with a To box, a Subject box, and a body, as shown in Figure 6-7. (As a reminder, the To and Subject boxes will say To and Subject in light gray text until you begin typing in them.) This screen looks and works a lot like the multimedia message editing screen in the Messaging application — just select the box you want to edit and start typing. As with other G1 applications where you need to enter text, you'll need to open the keyboard to get the job done.

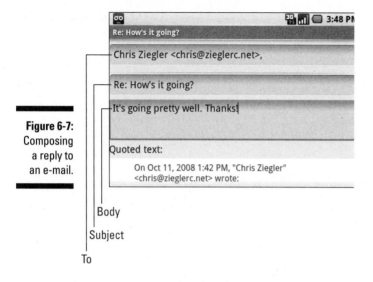

Figure 6-7:
Composing
a reply to
an e-mail.

Body

Subject

To

In the To box, separate each contact with a comma. When you begin typing, the G1 starts searching both the name and e-mail address fields of your contacts to try to find a match. If the G1 finds anything, it shows a list of matches below the box as you type. If you see your match, just touch it or select it with the trackball and it will be added to the list of recipients in the To box.

If you need to add carbon copy (CC) or blind carbon copy (BCC) recipients to the e-mail, the procedure is straightforward:

1. **Push the Menu button.**

2. **Touch the Add Cc/Bcc menu item.**

   Both CC and BCC boxes are added to the e-mail between the To and Subject boxes. They work exactly like the To box — just start typing a contact name or an e-mail address, and the G1 will attempt to locate a match for you so you don't have to type the whole thing. As before, separate multiple addresses with commas.

If you later decide you don't need the CC and BCC fields, deleting them is no biggie, either — just push Menu and touch Remove Cc/Bcc. (When you have the keyboard open, this may appear as "Remove Cc/..." because the G1 can't quite make the whole thing fit.)

When you've finished filling out the To, Subject, and body boxes, your e-mail is ready to send. If you're near the bottom of the e-mail, you'll see these three icons:

✔ **Send:** Immediately sends the e-mail off to its excited recipients.

✔ **Save as draft:** Saves your work to the Drafts view to be completed at a later time.

✔ **Discard:** Throws away the message you've composed.

Unlike deleting an e-mail you've received, discarding an e-mail you're in the middle of composing will not send it to the Trash where you can retrieve it later if you need it. A deleted e-mail is gone forever, so make sure you really want to get rid of it.

If you don't happen to be near the bottom of the e-mail, you won't see those icons, but never fear — there's no reason to scroll down to find 'em! Just push the Menu button and you'll have access to the Send and Discard commands from there. (You still need to use the icon at the bottom of the e-mail if you want to save your work to Drafts, however.)

Sending attachments

Unlike your PC, Gmail on the G1 will allow you to send only pictures as attachments, so if you need to send something else, such as a video or a track of music, you'll need to do it another way. (You could attach the G1 to your computer, transfer the files you want to send, and send the e-mail from the PC, for example. See Chapter 17 for details on how to connect the G1 to a computer.)

To attach pictures to an e-mail:

1. **While editing an e-mail, push the Menu button.**

2. **Touch the Attach menu item.**

   The Pictures application is called up, showing all pictures currently stored on your device.

3. **Select a picture.**

   You return to the e-mail editing screen, where you see a new section added between the Subject and body boxes. The new section shows a paper clip (it's an attachment, get it?), the file name of the attachment, its size, and an X icon (see Figure 6-8). Touching X removes the attachment.

4. **Repeat Steps 1–3 to attach additional pictures.**

Subject

To

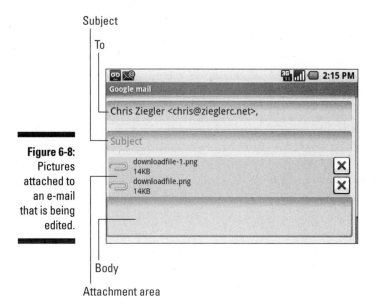

Body

Attachment area

Viewing received attachments

People probably e-mail you all sorts of attachments — pictures are just the tip of the iceberg. Maybe you get the occasional PDF file, Word document, or Excel spreadsheet. Even though the G1 doesn't have built-in applications for viewing or editing common document types, it has a neat trick up its sleeve: The G1 leans on Google to convert them into a format it can read.

This is how the process works: When you try to open an attachment that the G1 doesn't understand, it asks Google to convert it into HyperText Markup Language (HTML), which is the same code used to design and create Web pages. The G1 handles HTML like a pro, so once Google converts the document, you can view it right away. The most popular document types are supported, such as PDF, Microsoft Office, Rich Text Format (RTF), and OpenOffice.

When an e-mail you receive contains an attachment, you'll see a paper clip on its row in your Inbox, as shown in Figure 6-9.

Reading the e-mail as you normally would reveals one or more gray sections with paper clips to the left — these are your attachments. Picture attachments (see Figure 6-10) display Download and Preview icons; Download stores the picture on your G1 and shows it to you, but Preview simply shows it to you without downloading. Document attachments cannot be downloaded, so they display only a Preview icon (see Figure 6-11).

An attachment

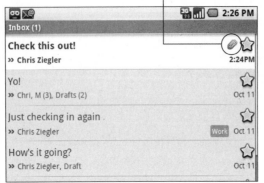

Figure 6-9: You've received an e-mail with an attachment. I wonder what it could be!

Figure 6-10: A picture attached to a received e-mail.

 Even though documents can't be saved to your G1, it's not a problem! As long as your phone is attached to a cellular or Wi-Fi network, you have access to Gmail — which in turn means you have access to all of these attachments by previewing them. Gmail's mantra of "don't delete e-mail; just archive it" definitely applies here.

 If you like a picture you've just received, you can set it as your wallpaper, a contact icon, or a myFaves icon in a jiffy. When you're looking at the image after touching Download or Preview, push the Menu button and touch Set as. You are presented with a new menu where you can choose what you'd like to do with the picture. (I get into these options in more detail in Chapter 11 when we look at Android's Pictures application.)

Figure 6-11:
An Excel
spreadsheet
attached to
a received
e-mail.

Gmail Settings screen

You access a handful of options for the Gmail application from its Settings screen. You can get to Gmail's Settings screen, shown in Figure 6-12, from the Inbox or any other label view by pushing the Menu button and touching Settings.

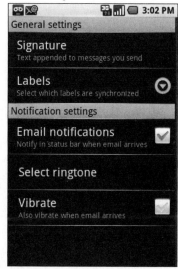

Figure 6-12:
The Gmail
Settings
screen.

The Setting screen is divided into two sections, General settings and Notification settings. Let's take a look at what's available in the General settings section:

✔ **Signature:** Touching this option brings up a small box where you can type some text that is added to the bottom of every e-mail you send from your G1.

A common signature people use is *Sent from my G1* or *Sent from my phone.* That way, your recipients are forgiving when you write an e-mail that's more terse than usual. (As good as the G1's keyboard is, you'd still rather not write a novel on it if you can avoid it.)

✔ **Labels:** This menu item lets you choose which labels' contents are automatically transferred to your G1. It's nice to have as much of your e-mail available on your phone as possible, but doing so is a trade-off: synchronizing more means more storage is required on the phone and more wireless data is used.

Fortunately, a compromise is available. When you're looking at the list of labels to be synchronized, touching one will toggle it between three settings, shown to the right of the label name: Sync all, Sync recent, and Don't sync (you won't see any text to the right of the label name when this option is selected). So if you have thousands of e-mails with your Work label, for example, you can choose the Sync recent option to avoid pulling down that mother lode to the phone.

E-mails that aren't synchronized can still be viewed and searched, but it just takes a bit longer because they need to be downloaded at the time you request them. They aren't downloaded ahead of time.

In the next section, Notification settings, you have three options:

✔ **Email notifications:** Like many other Android applications, Gmail gives you the option of disabling notifications received on your status bar. Unless you get so many e-mails that the notifications become useless, you probably want to leave this enabled.

✔ **Select ringtone:** You can select a tone to play when you receive an e-mail. Touching an option in the list of ringtones will play it once so you can preview it. Touch OK to confirm your selection or Cancel to revert to whatever you had previously.

✔ **Vibrate:** If this option is selected, the G1 vibrates when you receive an e-mail. Depending on how much e-mail you get, this could turn your phone into a nearly full-time buzzer!

The Email Application: Anything That Isn't Gmail

In a perfect G1 world, you'd use your Gmail account exclusively for sending and receiving e-mail, but for many of us, that's not a reasonable expectation. Perhaps you prefer the Web-based interface offered by another e-mail service such as Hotmail or Yahoo! Mail. Or maybe your job or school requires you to use another type of account. Whatever the reason, Android's Email application has you covered.

Specifically, Email supports any POP3- or IMAP-based e-mail server that can be accessed over the Internet. This covers an overwhelming majority of the world's e-mail servers. If your e-mail service has a Web site, go to it and look for a Help section on configuring e-mail clients (or talk to a customer service representative or administrator). Email can automatically configure popular e-mail services, but if you use a lesser-known or private service, you need to gather some details to use the application.

Setting up an e-mail account

To set up an e-mail account, begin by going to the Applications tab and choosing Email. The first time you start the Email application, you are prompted to set up an account, as shown in Figure 6-13.

Figure 6-13:
The Email application's welcome screen.

Enterprise e-mail, Android, and you

At the time of publication, Android doesn't support enterprise e-mail services such as Microsoft Exchange or BlackBerry Enterprise Server. However, the odds are very high that this will be resolved soon, either by Google or by any number of third-party companies offering add-on software. In the meantime, if you use one of these services to get your work e-mail, talk to your system administrator to see if he or she can forward your e-mail to another account such as your Gmail account. Although the sys administrator might not be willing to forward e-mail to another account because it's considered a security risk, asking is worth a shot.

1. **Touch Next to begin.**

2. **Enter your e-mail address and password where indicated (see Figure 6-14).**

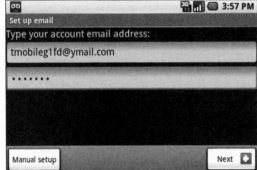

Figure 6-14: Enter your full e-mail address and password.

From this information, the G1 decides whether it knows how to automatically configure the account. If it does, you're in luck — no further setup is necessary and you can start using your account right away. If the G1 doesn't recognize the account type, however, you'll have to manually enter some details.

If you know that the application won't recognize your account type (if your company controls a private e-mail server, for example), you can prevent the phone from even trying to automatically configure the account by touching the Manual setup icon in the lower left of the screen.

For this example, I'm using a type of e-mail account that I know the G1 won't recognize: one of Yahoo Mail's lesser-used domains, ymail.com. That way, I can walk you through the process of setting up an account manually.

3. **Touch Next.**

4. **Choose the type of account, POP3 or IMAP.**

   IMAP is the more advanced type of e-mail server because it supports synchronization — if you read an e-mail on your PC, it appears as read on your G1 and vice versa. However, not all e-mail services support IMAP.

   I chose POP3 here, but the IMAP steps are basically the same.

5. **Fill in the incoming server settings:**

   a. **Fill in the incoming e-mail server's username and password.**

      At the top of the screen, the application has taken a guess at your incoming e-mail server's username and password based on the address and password you entered when you first started the setup process. If these are incorrect, you can change them here.

   b. **Enter the address of the incoming e-mail server and port.**

      In the next area of the screen, the application has also tried to guess the address of your incoming e-mail server and the port it should try to connect to. Because these addresses vary widely from service to service, odds are good that it guessed incorrectly, so pay close attention to these fields.

   c. **Choose a Security type.**

      Near the bottom of the screen, you're asked to choose the type of security. This is the type of encryption the application uses when connecting to your server. Encryption is almost always a good thing, especially for something as private and personal as your e-mail. If your service offers encryption, it's good to select it here. The type of security you choose depends on your service.

   d. **If you have a POP3 account, choose whether e-mail should ever be deleted from the server.**

      At the very bottom of the screen, if you are configuring a POP3 account, you'll be asked whether e-mail should ever be deleted from the server. If you choose Never, deleting an e-mail from your G1 will have no effect on your e-mail server — in other words, the e-mail will still be available to download to your PC or wherever else you may view it. If you choose After 7 days or When I delete from Inbox, be careful — you may encounter a situation where an e-mail gets downloaded only to your phone, and that can be a problem if you like to have all your e-mail available on your PC as well.

6. **Touch Next to continue the configuration.**

7. **Fill out the outgoing server settings:**

    a. **Enter the address of your outgoing e-mail server.**

    This is often, but not always, different from the address of your incoming server.

    b. **Below that, enter the port to connect to along with the security type, just as you did for the incoming server.**

    The application takes a stab at guessing the address and port, but as before, verify these carefully — they're quite likely incorrect.

    c. **At the bottom, specify whether the outgoing server requires sign-in, and if so, the username and password to use.**

8. **Touch Next.**

9. **Choose your account options.**

    Here, the application gives you an opportunity to set some basic options:

    - **Email checking frequency:** Because the Email application doesn't support the same push technology that the Gmail application does, you can configure Email to poll your server on a regular basis to check for new e-mail. This option is set to Never by default, meaning the G1 will not check for e-mail on this account unless you explicitly ask it to. You can change this setting so that e-mail is automatically retrieved at intervals ranging from every 5 minutes to once per hour.

      Choosing a frequent interval (such as 5 minutes) increases battery drain, which means you'll have less use of your G1 between charges. Select a less frequent setting for better battery life; to really maximize battery life, choose Never and just check for new e-mails manually.

    - **Send email from this account by default:** If you select this option, the G1 automatically uses this e-mail account whenever you send an e-mail from elsewhere in the phone (from contacts, for example).

    - **Notify me when email arrives:** This option functions similarly to the notification feature for Gmail. When selected, you'll receive notification in the status bar that you have new e-mail (but only if you have your Email checking frequency set to something besides Never).

    Don't fret over these options too much — you can always change them later.

10. **Touch Next.**

11. **If you want, enter an optional name for the account.**

    The optional name is useful if you intend to set up multiple e-mail accounts.

12. **Enter the name you want attached to e-mails you send.**

    By default, your first and last name is used here, but you're free to change it.

13. **Touch Done.**

    You are taken to your Inbox for this account.

Retrieving new e-mail

If you have the Email application set up to automatically retrieve e-mail at a particular interval, you'll start to receive notifications in the status bar for this account (unless you've disabled the notifications). Figure 6-15 shows a notification from the Email application. In this example, the notification is for a Gmail account that I have set up in the Email application — even though a dedicated Gmail application is available, nothing is stopping you from configuring Gmail accounts in the Email application as well! This capability is particularly handy if you use multiple Gmail accounts, since the Gmail application supports only one.

Figure 6-15:
A new e-mail notification from Android's Email application.

Note that when an e-mail arrives, Gmail shows an @ symbol in the status bar but Email shows an envelope with a small @ symbol above it and to the right. This helps you distinguish at a glance whether the new e-mail is from Gmail or another e-mail account.

Touching the notification takes you to your Inbox (see Figure 6-16). This is one way to start the Email application, but you can start it also from the Applications tab; now that you have an account set up, the application will take you straight to your Inbox instead of prompting you to set up a new account.

Figure 6-16:
The Email application's Inbox screen.

The display here is similar to, but not exactly the same as, the Gmail application's Inbox. Unread emails have a thin vertical green bar to their left and a bold typeface.

You can retrieve new e-mail from your Inbox by pushing the Menu button and choosing Refresh.

About folders

You'll notice that the Inbox row is at the top level while e-mails below it are indented. That's because these e-mails are *within* the Inbox folder. Unlike the Gmail application, which presents different labels in different screens altogether, the Email application presents them all in one screen as separate folders. If you're using an IMAP account, you may see several folders here in addition to the Inbox — these were automatically pulled down from your server, and you may recognize them as the same folders you see when you access this account from your PC.

You can expand and collapse these folders by touching them. If you start an e-mail and save it as a draft, a Drafts folder is created. And if you delete an e-mail, a Trash folder is created.

Reading and working with e-mail

The process for reading and managing your e-mail in the Email application is similar to the process in Gmail, except for a few missing options specific to Gmail, such as Change labels and Mark as spam. At the bottom of each e-mail, you still have access to the Reply, Reply all, and Delete icons, which work as you'd expect. You can also push the Menu button from within an e-mail to Reply, Reply to all, Forward, Delete, or Mark an e-mail as unread.

If you're looking at the Inbox view, you can touch and hold on an e-mail to call up a menu of actions that you can perform on that e-mail:

- ✓ **Open:** This option simply opens the e-mail — it's no different from touching an e-mail in the Inbox list.

- ✓ **Delete:** The Delete option sends the e-mail to the Trash folder and removes it from the Inbox.

- ✓ **Forward, Reply all,** and **Reply:** These items create new e-mails based on the e-mail you've selected — they work as you'd expect.

- ✓ **Mark as unread/Mark as read:** If you've selected a read e-mail, this will mark it as unread, and vice versa.

These options are pretty much the same as the menu items you see when you push Menu from within an e-mail itself (see Figure 6-17).

Figure 6-17:
Menu items available when touching and holding on an e-mail.

Viewing attachments

Gmail has the luxury of relying on Google's servers to turn document attachments into HTML that can be processed and displayed by your G1, but attachment viewing options are much more limited in the Email application. If you come across an e-mail with a document attachment, it will look something like Figure 6-18.

Figure 6-18: An attachment that can't be viewed or saved. Bummer!

You can see the attachment's name and size, but you can't view it — you'll need to return to desktop-computer-land for that.

On the other hand, picture attachments can still be viewed and saved. When you get one, you'll see Open and Save icons to the right of the attachment — these icons work the same as the Preview and Download icons, respectively, when looking at a picture attachment in Gmail.

Writing and sending e-mails

The process of writing and sending e-mails from the Email application works pretty much like the process in the Gmail application, which is nice since you won't have to remember two ways to go about things if you find yourself using both apps.

To compose a new e-mail from scratch, push the Menu button from the Inbox view and touch Compose. Otherwise, you can reply to or forward an existing e-mail, but in any case, you'll arrive at a screen that looks like Figure 6-19.

Figure 6-19:
Composing
a new
e-mail.

All the same rules from Gmail apply here. You can send this e-mail to multiple recipients by separating them with commas, and as you type new recipients, your G1 will attempt to match your typing to names or addresses stored in your contacts — just touch a match to select it and the rest of the address will be filled out for you.

By pushing the Menu button from this screen, you can add CC and BCC boxes or add picture attachments (the menu item may appear as "Add attachm…", but you get the idea). You can also send the message, save it as a draft (in which case it'll appear in your Drafts folder from the Inbox view), or discard your new message. These final three options are also available as icons at the bottom of the message, so there's no need to push Menu to access them as long as they're visible on the screen.

Working with multiple accounts

Since Email can support multiple accounts at the same time, there has to be a good way to manage them, right? Indeed, there is — and it's just one menu item away. When you're looking at the Inbox view for a particular e-mail account, push the Menu button and choose Accounts. You'll see a screen that looks like Figure 6-20.

By touching and holding a particular account in this list, you'll be presented with three menu items:

✔ **Open:** This menu option returns you to the Inbox view for the selected account. It performs the same function as simply touching the account name.

✔ **Account settings:** This option lets you modify the settings you configured when you first went through the setup for this account, such as the
name to display when sending e-mails from this account and the incoming and outgoing server information. The Account settings screen can be
accessed also from the Inbox screen for a particular account by pushing
Menu and choosing Account settings.

✔ **Remove account:** This option deletes the selected account from your
list of accounts. You won't be able to access it unless you set it up again.

Figure 6-20:
Your e-mail
accounts.

By pushing the Menu button from the account list, you can refresh (that is,
get new e-mail for) all the listed accounts, add a new account, or compose a
new e-mail. If you choose Compose, the e-mail account that you have specified as your default is used to send the message.

Chapter 7

Navigating with Google Maps

*I*nside your G1 are three amazing pieces of hardware (well, way more than three amazing pieces of hardware are in there, but three are of particular interest to us in this chapter): a GPS receiver, a magnetic compass, and a three-axis accelerometer. Combined, these guts make the G1 one of the most powerful phones ever to support navigation. Let's look at these three components in a little more detail.

✔ **GPS receiver:** The GPS receiver listens to signals from GPS (Global Positioning System), an extensive constellation of satellites covering the globe that devices can use to determine your precise coordinates anywhere in the world. If you have a navigation system in your car or a handheld navigation system made by a company such as Garmin or Magellan, or you've been on an airliner in the past couple of decades, you're well-acquainted with the benefits of GPS!

✔ **Magnetic compass:** There's a problem with GPS, though — it can determine only where you are, not what direction you're facing. (Some GPS devices can figure out your direction by measuring the direction traveled between two nearby locations, but that works only when you're moving, not when you're standing still.) The G1's magnetic compass is truly unprecedented for a phone of any kind. Several other phones offer compasses, but there's a big difference: Unlike other models, the G1 allows makers of applications to tap into the data in the compass and use it. The G1's compass allows the phone to determine where it's pointed, and applications (like Maps, for example) can use that directional information to do cool things.

✓ **Accelerometer:** The final piece of the puzzle is the G1's three-axis accelerometer. "Three-axis accelerometer" is quite a mouthful but means one simple thing: The G1 can figure out whether you are holding it with the screen facing up, down, left, right, or at a strange angle. The accelerometer comes in handy in the Maps application — you'll see why a little later in this chapter — and it also lets you control some downloadable applications and games just by moving your G1 around!

Many of us have used Google Maps on our PCs at one time or another to find an address on a map, get directions from one location to another, or just play around with its super-cool satellite and street-level views. The three technologies I just described — the GPS, compass, and accelerometer — come together to make Maps on the G1 even cooler than on your PC because the G1 can use the information from these technologies to determine your current situation and integrate it into what you're seeing on the map right now.

In this chapter, I show you how to make absolutely, positively sure you're never lost again (well, as long as you have your G1 with you) by taking full advantage of Android's integrated Maps application.

Starting Maps

Normally, you start the Maps application from its icon in the Applications tab or a shortcut on the Home screen (if you've created one), but you can also start it directly from a postal address on a Web page in the Browser or a contact's details screen in the Contacts tab of the Dialer application. If you have assigned a postal address to a contact, you can open Maps and immediately show the address on the map by simply touching it — a handy way to quickly see the address of a business or home that you're traveling to. (And yes, Maps can tell you exactly how to get there, too! More on that later in the chapter.)

Getting Around Maps

When you first open Maps, you see a screen similar to Figure 7-1. If you've ever used Google Maps on your PC, this view should be familiar; it's simply Google's rendered map of streets, highways, and other navigational aids.

Figure 7-1:
The Maps
application.

Scrolling the map

To scroll around the map, you can press and hold on the screen and then move your finger; the map moves as your finger moves. If you prefer to navigate with the trackball (this is especially convenient if you have the keyboard open), just roll the trackball and the map moves.

While using your finger, you can also flick the map. To do this, make sure your finger is still moving along the map when it loses contact with the screen, and you'll see that the map continues to move for a while in the direction your finger was moving. After a moment, the map decelerates and eventually comes to a stop. (Flicking sounds complicated, but you'll find that it becomes a natural motion once you've tried it a few times.)

As you scroll around the map, you might find that new areas that have just scrolled onto the screen are displayed as a checkerboard pattern (see Figure 7-2) that is eventually replaced by the map you're expecting to see. This happens because the G1 doesn't keep all map information in its memory — that would take *way* too much space! Instead, the Maps application asks Google for the new parts of the map that are now needed to display. Depending on the speed of your data connection, this display can take anywhere from less than a second to as much as ten seconds or more.

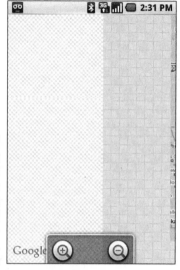

Figure 7-2:
Parts of
the map
have not
yet finished
down-
loading.

Zooming

Depending on what you're doing, you'll want to see the map at a different level of zoom. For example, if you're looking at New York City and you need to see San Francisco, you won't flick the screen to the left at the same zoom level you were using to get an up-close look at Manhattan (unless you have an entire day to kill). Instead, you'd want to zoom out a whole bunch, move over to northern California, and zoom back in.

Zooming in Maps is a cinch, and you can do it in several ways. By touching the screen while the map is showing, you bring up zoom controls at the bottom (see Figure 7-3) that you can touch to zoom in and out. The magnifying glass with the plus sign zooms in one level, and the magnifying glass with the minus sign zooms out one level.

You can call up the zoom controls also by pushing Menu, selecting the More menu item, and selecting Zoom from the list that appears. Finally, you can avoid touching the screen at all: Press the Menu key and the Z key at the same time, move the trackball left and right to select the zoom in or zoom out command, respectively, and then press the trackball to execute the zoom.

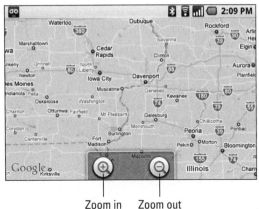

Figure 7-3:
The Map
applica-
tion's zoom
controls.

Zoom in   Zoom out

Four Maps, One Application

The default map that you see when you open Maps is convenient, and it's often the easiest way to see streets and other navigational features when you're walking or driving to a destination. However, the Maps application supports four — count 'em four — map modes.

To choose a map mode, push the Menu button while looking at a map and select the Map mode menu item. You are presented with the list of map modes; just select one to switch to it. Let's take a closer look at what each of these modes does, how they look, how they behave, and when you might use them.

Map mode

The default map mode, which you've already seen in Figures 7-1 and 7-3, is known simply as Map. It's as good of a description as any because this is the most useful mode and will likely be the one that you spend the most time using. The Map mode is best equated to the style of map you might carry with you in your car's glove box — it highlights roads, parks, airports, and other major navigational features without any unnecessary clutter.

As you zoom in and out in this mode, Maps increases and decreases the level of detail to the appropriate amount for the current zoom level; for example, if you're zoomed all the way out so that you're looking at the entire United

States on the screen, you'll see nothing but state names, borders, major cities, large bodies of water, and a few national parks (indicated in green). If you zoom in so that you're looking at a single block in Chicago, though, you'll see individual street names, the directions of one-way streets, subway stations, and the shapes of buildings.

Satellite mode

The Satellite map mode (see Figure 7-4) uses Google's extensive collection of satellite imagery to supplement the standard Map mode. Instead of seeing nondescript gray and white areas around streets, you see buildings and terrain as snapped by photo satellites circling the globe. (It's amazing that they can get such great shots from way, way up there, isn't it?) In this mode, streets and street names are overlaid with a little bit of translucency on the imagery so that they don't obstruct the terrain underneath.

Figure 7-4:
The Satellite map mode.

This mode can be cluttered for day-to-day navigation use, but sometimes it can be convenient (or just plain cool) to be able to see landmarks near the area you're looking for.

Because different parts of the world were photographed by different satellites (or, in some cases, by the same satellite but with different levels of zoom), not every corner of the globe is available in Satellite mode at the closest zoom levels. If you zoom in beyond what Maps can provide, you see a screen of X symbols above a checkerboard, as in Figure 7-5. Just zoom back out a level or two to continue on your virtual journey.

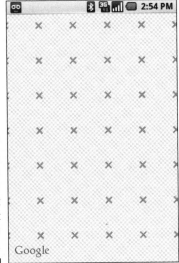

Figure 7-5:
The area
you're look-
ing at isn't
available at
this zoom
level.

Traffic mode

Another fantastic feature of Google Maps that you may already be used to from using it on your PC is the ability to show traffic conditions on major highways and byways. Well, good news: That same capability is available to you right from the comfort and convenience of your G1.

The Traffic map mode (see Figure 7-6) is essentially the same as the Map map mode, with one important distinction — roads for which Google can get traffic information are displayed as color-coded superimposed lines. The information comes from a variety of sources, which Google aggregates and updates every few minutes. These roads are shown in any of the following colors:

- **Green:** Traffic is moving at 50 miles per hour or better.

- **Yellow:** Traffic is slowed, moping along at somewhere between 25 and 50 miles per hour.

- **Red:** Ouch! Things are jammed here, with average speeds below 25 miles per hour.

- **Gray:** Although Google usually receives traffic data for this road, no cur-rent data is available. Don't blame Maps if you travel to this road and discover that the traffic has turned it into a parking lot!

Figure 7-6:
The Traffic
map mode
shows
colored
areas of two
express-
ways that
provide
Google
with traffic
information.

Street View mode

If you've been wondering where the G1's magnetic compass and accelerometer come into play, here's your answer. Street View takes you right down to a perspective view of the location you're looking at on the map — just as though you were standing there! To make this happen, Google has taken the incredible step of deploying fleets of camera-equipped cars around the United States to snap pictures everywhere they go. As you can imagine, this is a massive, time-consuming undertaking. At the time of this writing, Street View was available in 69 metropolitan areas of the United States, but that number is growing and existing Street Views are frequently updated with fresh pictures.

Enabling Street View and taking full advantage of it takes a few steps:

1. **Choose Street View from the Map mode menu.**

   You see a map that looks essentially the same as the Map map mode, but streets that have Street View enabled are highlighted with a blue outline.

2. **Find a location that you'd like to see at street level, and tap twice on the screen in that location.**

You see a message that the Maps application is searching for a Street View for that location. After a moment, the address of the location you've tapped appears, along with a message to "Press here to enter Street View," as shown in Figure 7-7.

Figure 7-7: A Street View is available at this location.

3. **Touch the message where the address appears.**

This is where the magic starts to happen. You are transported to a street-level view of the location you tapped (see Figure 7-8). Translucent lines indicate streets that can be navigated in Street View mode. By touching an arrow on any one of the lines, you are moved in that direction, facing the same direction as the arrow. As you encounter streets, you can touch their labels to move to a Street View on that street. You can also zoom the Street View by touching (or selecting with the trackball) the magnifying glass icons at the bottom of the screen.

At this point, you have two ways to navigate around Street View. By default, you can look up, down, and around by moving your finger around on the screen or by scrolling the trackball. That's not the cool way, though! By enabling Compass mode, the G1 will use its compass and accelerometer data to automatically position the Street View based on the direction you're facing and the way you're holding the phone. Here's how to enable Compass mode:

1. **While in Street View, push the Menu button.**

2. **Select the Compass mode menu item.**

A brief message indicates that the Compass mode is now on.

3. **Move in a circle and tilt the G1 up and down.**

You'll notice that the view responds to your movements — you see exactly what you'd be seeing if you were standing at that location!

4. **To stop the G1 from responding to your movements and return to traditional trackball navigation of Street View, repeat Steps 1 and 2.**

Figure 7-8: Street View, showing an intersection and touchable street labels.

If you're trying to find an unfamiliar location, you can use Street View to "visualize" your destination before you even get there. Just go to Street View for the address you're looking for, enable Compass mode, and walk around to get a real sense of the area before you arrive; that way, you'll be familiar with landmarks ahead of time.

If you see an inappropriate or incorrect image while browsing Street View (hey, it happens — we're talking about millions of images of real locations!), push the Menu button and select the Report image button. You'll be taken to a Web page in the Browser that lets you tell Google about the problem.

To return to the map from Street View, just push the Back button or push Menu and select the Go to map item.

While you're looking at the map, you can drag the Street View man icon around to choose a new location to view — just touch his icon and move your finger around the screen (see Figure 7-9). Wherever you let go is where the Maps application will look for a Street View next.

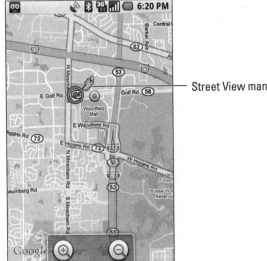

Street View man

Figure 7-9:
Dragging
the Street
View man
icon around
the map.

Finding Where You Are

I told you at the beginning of the chapter that as long as you had your G1 with you, you'd never be lost again — but so far, nothing you've done in the Maps application has involved your location. You've just scrolled through the different map modes, looking at interesting buildings (and in the case of Street View, maybe an interesting person or two), but none of this cool stuff necessarily tells you where you're currently standing.

So how do you find out where the heck you are? It's a piece of cake! From any map mode, push the Menu button and select the My Location menu item. After a moment, the map moves so that your current location is in the center, represented by a blue dot.

You may notice a light blue circle around the blue dot (see Figure 7-10). This happens when the G1 is unable to determine your precise location but *can* say with reasonable certainty that you're somewhere within the blue tinted area.

Assisted GPS: The G1's secret sauce for getting around poor GPS reception

Why does the tinted blue circle sometimes appear after selecting My Location? Why doesn't the G1 always know exactly where you are? GPS satellites require a direct, unobstructed, line-of-sight view. Reception can be impossible if you're indoors or surrounded by tall buildings. (In fact, GPS was never designed to work with cellphones, so it's impressive that GPS works as well as it does.) You generally get the blue tint because you don't currently have great (or any) reception with GPS satellites. To make up for a lack of satellite reception, the G1 makes use of a combination of technologies that are together known as aGPS (assisted GPS). aGPS combines traditional GPS reception with a known database of Wi-Fi and cellular tower locations to help the G1 take an educated guess at your approximate location. You'll be surprised at just how accurate that guess usually is.

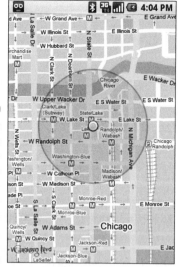

Figure 7-10:
The Maps application knows that you're somewhere within the blue tinted circle.

By default, the GPS hardware in the G1 is turned off to conserve battery life. The Maps application can usually still determine your location but with less accuracy. GPS is enabled using the Settings application, which I cover in Chapter 15.

What can you do with your location information once the Maps application has it? Touching the blue dot reveals the Details screen shown in Figure 7-11.

The Details screen lists your exact coordinates and the following options:

- ✔ **Zoom in to My Location:** Takes you all the way down to the closest zoom level available on the map and centers it on your location.

- ✔ **Directions to this address:** Takes you to the Directions screen and automatically enters your location as the destination. (I discuss directions in detail later in the "Getting Directions" section.)

  The Directions to this address option is handy for guiding someone on the phone to your current location — just get the person's current address and use it as the starting location.

- ✔ **Directions from this address:** Takes you to the Directions screen and enters your location as the starting address.

- ✔ **Edit My Location settings:** Opens the My Location settings in the Settings application, which we look at in Chapter 15.

- ✔ **Disable My Location:** Removes the blue dot from the map. (You can always get it back by pushing Menu while looking at the map and choosing the My Location menu item.)

Figure 7-11:
The My
Location
Details
screen.

Searching for Locations

The Maps application is every bit as adept at searching for business names, addresses, points of interest, and categories of locations as its PC equivalent. It's not a stretch to say that you have the power of every Yellow Pages and White Pages in the world stuffed into an object that slips into your pocket, which is pretty amazing.

To perform a search for locations:

1. **Push the Menu button and select the Search menu item.**

   Alternatively, you can just press the Search key. Or even better, just start typing. Maps will know that you're trying to conduct a search.

   A search bar appears at the top of the screen (see Figure 7-12). Recent searches appear in a list directly below the text box. If you want to repeat one of these searches, you can just touch it and the search begins immediately. Otherwise, continue to Step 2.

Search bar

Figure 7-12:
The Maps application's search bar.

Recent searches

2. **Begin typing your search.**

   Your search can be in the form of a street address, a city name, a business name, a category (such as "pizza"), or a contact name for which you've entered one or more street addresses. As you type, your search history starts to be filtered so that only items that match what you're entering are visible. You also see any contact names with street addresses that match your search. If you see what you're searching for, touching it will start the search.

3. **When you've finished typing your search, press the Enter key or touch the Search icon on the screen.**

Viewing search results as a list

The search you've run in the last section takes a moment to finish as the Maps application asks Google for information. When the query is complete, if there's only one possible match, you are taken directly to it in the Map view; otherwise, you are presented with a list of possible matches (see Figure 7-13). Generally, Maps looks for matches in the vicinity of wherever you happened to be looking on the map when you ran the search, but if you typed a specific address or city, it does its best to look in the correct area.

Figure 7-13:
A list of possible search matches.

At this point, you can touch a match to show it (and center it) on the map, or you can touch and hold on a match for a moment to get a pop-up menu. This menu has the following options:

✓ **Zoom in to address:** The first option zooms the map all the way in and centers the selected match. This option is the same as simply touching the match's row.

✓ **Directions to here/directions from here:** This option takes you to the Directions screen and enters the selected match as the destination address or starting address, respectively.

✔ **Add as contact:** Because Google Maps is frequently able to gather the full name, street address, and phone number of the search matches that it returns, touching this menu item conveniently turns the match into a new contact in your contacts list.

Viewing search results on the map

At the bottom of your search results list are two buttons. Touching the button on the left, Edit search, simply allows you to tweak your search terms and run the search again (remember, *pizza* has two *z*'s). The button on the right, Show map, takes you back to your map view with all the matches in the list added as pushpins (see Figure 7-14).

Previous search match

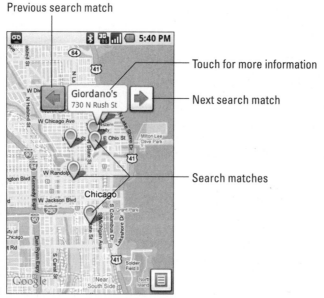

Touch for more information

Next search match

Search matches

Figure 7-14:
Viewing
your search
results on
the map.

To choose a specific match, touch its pushpin — an information bubble appears above it with the match's name and street address. You can also navigate sequentially from one match to the next by using the right and left arrows that appear on either side of the information bubble.

Once you've located the match that you're interested in, touching its information bubble will take you a Details screen. (You may remember a similar Details screen when touching the blue My Location dot.) In here, you see the match's full name, address, city, and state, along with a host of options. You can use these options to zoom all the way in to view the match on the map, call its primary phone number, use it as the destination or starting point to get directions, add it as a contact for quick dialing in the future, and — if available — open the Browser to visit its Web site.

Getting Directions

The true beauty of the Maps application isn't just that it shows you the location of things (including you) — it's that it'll show you how to get there, too. In this chapter, you've seen several ways that the locations you work with in Maps can be used as the starting points or destinations for getting turn-by-turn directions. In this section, I show you exactly how that feature works.

If you select the Directions to here or Directions from here menu item that you've seen elsewhere in Maps, you're already at the Directions screen (see Figure 7-15) with a location entered into one of its two fields. Otherwise, you can push the Menu button and select the Directions menu item to get here.

Fill in text box using current location,
contact address, or Maps history search term

Figure 7-15: The Directions screen.

Swap starting point   Destination   Starting point
and destination   text box   text box

At the bottom of the screen are two text boxes; the top one is your starting point, and the bottom one is your destination. These boxes work a lot like a location search does: You can enter a street address, a business or city name, a contact name, an airport code (such as ORD or SFO), and so on. As you type, Maps looks through your search history and your contacts, and if it finds any potential matches, it offers them to you in a list. You can either select an item from the list or continue typing.

To the left of the text boxes, you'll notice a button with up and down arrows. Touching this will simply swap the contents of top and bottom boxes, reversing the starting point and destination.

Touching the button to the right of the text boxes will allow you to select your current location, an address from your contacts, or a search term from your Maps history to fill out the boxes. This feature is a nice little shortcut to avoid entering text unnecessarily.

When you have filled out the text boxes, touch the Route button on the far right side. Google calculates the directions for you and, after a moment, displays a list of turn-by-turn directions to reach your destination (see Figure 7-16).

Figure 7-16:
Turn-by-turn directions from San Francisco to New York — one heck of a drive.

The first and last items in the list are your starting point and destination, respectively. In the middle you have each turn (or other driving instruction) that you need to follow. On the right, the distance before the following instruction is listed. At the very bottom, you see the total distance and estimated drive time (subject to bathroom breaks, of course).

To see your route highlighted in blue on the map, touch the Show map button at the bottom of the screen. Alternatively, you can touch a specific instruction to see the location of that instruction zoomed in and centered on the map. Once you're looking at the map, each turn point is highlighted by a diamond in a white circle; touching one of these points will show the textual instructions that accompany it inside a bubble (see Figure 7-17). You can move backward and forward through the instructions by touching the left and right arrows, respectively, or you can return to the list of directions by touching the icon in the lower-right corner of the screen.

Although the Directions feature is designed mainly for car use, note that unlike an in-car navigation system, the Maps application will not automatically advance from one instruction to the next as you drive (you have to do that yourself). In addition, the app will not speak the instructions to you — you need to look down at the screen to see what's going on. Obviously, these activities are not safe for a driver who's trying to pilot a car at roadway speeds. Please, for your own sake (and for mine, if I'm ever near you on the road), let a copilot perform these actions for you!

When you're finished with your route, you can push the Menu button and select the Clear map menu item to remove the route from the screen and return to the normal map mode.

Previous instruction   Next instruction

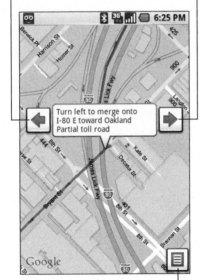

Figure 7-17:
A turn point
for your
grand jour-
ney from
San Fran to
New York
highlighted
on the map.

Return to the list of directions

If there's one G1 application that turns you into a superhero, it's Maps — you'll never be lost, you'll always know where to find what you're looking for, and if you're feeling generous, you'll be the one who can bail out that helpless looking tourist who can't find the highway. And, hey — isn't it a good feeling to know that you'll be able to find a Vietnamese restaurant the next time you're in Dubuque, Iowa? (Hint: It's the Asian Gourmet.)

Chapter 8

Contacts

*W*hen you turned on your G1 for the very first time and entered your Google account information, something magical happened: All the contacts stored in your Google account were downloaded automatically. Even if you weren't actively using Google's contacts feature before, as long as you were using Gmail, Google was quietly collecting the names and e-mail addresses of people you e-mailed (and who e-mailed you) and filing them away in your contacts list. Granted, that's not all the information you need about the people you need to contact — and odds are lots of people are in there who you'll never be contacting again — but it's a start!

Even if you didn't have a Google account previously or weren't using Gmail, creating an account and actively using your G1 puts you on a fabulous path to address book nirvana. Why? Unlike many cellphones that don't give you a nice, clean way of getting your contacts into their tiny little brains, all contacts you create on your G1 are automatically and silently synchronized to your Google account. That means two very cool things: one, the contacts are available on your PC simply by going to www.gmail.com and logging in, and two, any Android phone you ever buy in the future will be just one Google account login away from having all your contacts. How cool is that?

I've already touched on some features of the Contacts system in the G1 in Chapter 4, but now we're really going to get into the nitty-gritty of what makes it tick.

Contacts Is an Application but Not Really

The Contacts application in Android has a dirty little secret: It's not really its own application. Contacts is nothing more than a tab in the Dialer application (see Figure 8-1), which is a logical location for it considering that its primary function is as a repository of phone numbers.

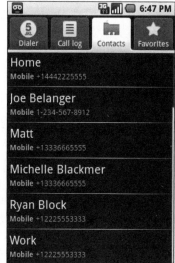

Figure 8-1:
The
Contacts
tab.

That said, Contacts is still treated like its own app in some respects. It gets its own icon in the Applications tab — in fact, it's one of the default icons on the G1's Home screen when you first turn on the phone. Choosing this icon means you'll be taken straight to the Contacts tab in the Dialer, but you can get to it also by opening the Dialer application (or pushing the Send button) and touching the Contacts tab.

Finding Contacts

They say the world has two kinds of people: those who have a ton of contacts in their phone and those who are going to. Whether you're in the first category or the second, it's a good idea to get a handle on how to find a contact quickly and efficiently, and Android gives you a few neat tools to do just that.

You can just scroll through your entire list using swipes of your finger or the trackball, but once your contact count gets much above a hundred or two, you'll find that this gets old *really* fast. So let's have a look at how Android can help us.

The first method uses a tab to seek out a particular letter, just like those old-fashioned Rolodexes (don't worry, this method is a whole lot more portable and won't make you look like you're stuck in the 1980s, I promise):

1. **Start scrolling through the list of contacts, either with the trackball or your finger.**

   If you have more than one page's worth of contacts, a gray tab appears on the right side of the display.

2. **Place your finger on the gray tab and move it up or down.**

   A letter appears in the center of the screen (see Figure 8-2) — contacts that start with this letter are currently scrolled into view. By moving the gray tab up and down, you can move through the alphabet one letter at a time, which has the effect of rapidly scrolling through your contacts.

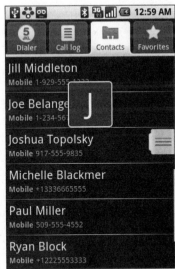

Figure 8-2:
Scrolling
through
your
contacts,
Rolodex-
style.

3. **Once you get to the letter that your contact starts with, let go of the gray tab.**

   After a moment, the tab disappears. (You can make it reappear by scrolling the screen as you did in Step 1.)

4. **To fine-tune your search and get to the specific contact you're looking for, scroll without the gray tab using either your finger or the trackball.**

The second method is more direct but requires that your keyboard be open:

1. **With the Contacts tab open, start typing either the first or last name of your contact.**

   The letters you're typing appear in the middle of the screen, as shown in Figure 8-3.

Figure 8-3:
Typing a contact's name to find the contact in the list.

2. **Keep typing the name until the contact you're looking for is visible.**

   Once you see it, you can select it as usual.

But wait — there's one more trick for finding a contact with ease. In Google, your contacts are organized into groups, and the default group is called My Contacts. If you used Gmail before you used Android, Google automatically started adding frequently used contacts to your My Contacts group, so if you e-mail a lot of people, this group can get really big *really* fast.

Using the Gmail Web site on your PC, you can create new groups and add contacts to them; for example, you might want to have a Work group and a Personal group. How you name and organize these groups is up to you.

Once you have those groups set up to your liking, they automatically appear on your phone; you don't have to do anything. Then you can configure your Contacts tab to show only the contacts from a certain group. Here's how:

1. **While in the Contacts tab, push the Menu button.**

2. **Touch the Display group menu item.**

You're looking at a menu of available groups. In addition to My Contacts and the groups you've created and organized through the Gmail Web site, you'll have a Contacts with phone numbers option — handy for quickly seeing only contacts that you can dial, considering that Gmail automatically adds new contacts to your list with an e-mail address alone.

3. Choose a group to display by touching it, and then touch OK.

That's it! Your screen looks a lot cleaner now, doesn't it? To return to your full list of contacts, simply repeat Steps 1 through 3, touching the My Contacts item from the menu.

Creating a Contact

In Chapter 4, I mention that you can start the process of creating a contact from the Dialer or Call log tab. You create a contact from the Dialer by entering a number, pushing Menu, and touching Add to contacts; from the Call log, you touch and hold an entry until the pop-up menu appears and then you touch Add to contacts.

These are convenient methods — particularly from the Call log function, where you may get a call from someone who's not in your contacts and immediately realize that you'd like to add him or her. Of course, you can also create a new contact from scratch:

1. From the Contacts tab, push the Menu button.

2. Touch New contact.

Regardless of whether you've created this contact from the Dialer tab, the Call log tab, or the Contacts tab, you arrive at the same screen, which is shown in Figure 8-4. The only difference is whether a phone number has been filled out for you.

Figure 8-4: The contact editing screen.

Let's take a look at the fields on this screen:

- **Add icon:** By touching Add icon, you can assign to your contact a picture that appears every time the contact calls. You'll be taken to the G1's picture viewer, and this is where it gets cool! When you select a picture from this screen, Android attempts to find and automatically highlight any faces that it finds in the picture, as in Figure 8-5 (don't ask me how it knows — it's a mystery).

Figure 8-5:
Android
attempts
to find the
faces in this
picture —
one out of
two ain't
bad.

If the picture viewer can detect faces, it places a box around each one that it finds; if it can't find any faces, it puts a single highlight box in the middle of the picture. Either way, you touch the box you want to use to select it, and this serves as your crop box for the picture. Dragging your finger within the box moves it around the image. Dragging your finger on the box's border changes the size of the box. The part of the picture within the box becomes the icon for this contact, so crop carefully! Once you're satisfied with your work, touch Save (or touch Discard to cancel and return to the editing screen without making any changes to the contact's icon).

- **First and Last box:** This is where you enter your contact's first and last name.

It's important that you maintain a consistent first, last ordering here — you don't want a mix of Joe Smiths and Smith Joes in your contacts list because they would end up being listed out of order and it would become difficult to find them by scrolling.

- **Dial number fields:** The area under the Dial number heading lists one or more phone numbers and number types.

Number types can be common terms such as Home, Mobile, Work, and Fax, but you can also add your own. The default number type for a new number is Mobile. To change the default type, touch it and a menu appears; touch a type name in the list to select it. To use your own type name, scroll to the bottom of the menu and touch Custom. A new window appears, prompting you to type a new name; touch OK when you're finished.

Once you've selected a number type, highlight the Phone number box and type the phone number.

✔ **Send email fields:** These fields look — and work — much the same way as the Dial number fields, except you're entering an e-mail address instead of a phone number.

The default e-mail address type is Home. To change the default, touch it and a menu appears. Just like your phone numbers, you can create custom types by scrolling to the bottom of this menu and touching Custom.

✔ **Ringtone:** By default, all contacts sound the same ringtone when they call — but where's the fun in that? Selecting a custom ringtone for a contact is an easy way to know when that special someone is calling without even having to glance at your G1. Touch the box to the right of the Ringtone label to select a different ringtone; touching a ringtone name plays it so you can preview it, and touching OK assigns it.

✔ **Send calls directly to voicemail:** Speaking of "special someone," we all have a few people in our lives who we'd rather not talk to from time to time. With Android, your phone can do the dirty work of sending the call straight to your voice mail box for you. When this option is selected and the contact calls, you'll never even hear your phone ring (or vibrate, as the case may be). If you're looking to send *all* your calls to voicemail, there's a simple solution for that, too — turn off your phone!

✔ **More info:** You can add a lot of additional information to a contact besides a picture, a name, a phone number, and an e-mail address, and this is where you do it. Touching the More info icon displays a menu of various fields you can add to this contact.

If you want to add additional phone numbers and e-mail addresses — say, one each for home and work — you can do that by touching the Phone or Email items, which open submenus of available phone number and e-mail types, respectively. As before, you can scroll down and touch Custom to enter your own.

Additionally, you can add the contact's various instant messaging (IM) accounts, street address, company, and job title (grouped under Other as the Organization option), as well as freehand notes.

After you've finished entering information for this new contact, touch Save at the bottom of the screen, or touch Discard changes to cancel.

Doing Stuff with Contacts

It turns out that contacts aren't just for making phone calls anymore! Every type of information that you add to a contact — phone number, e-mail address, instant message name, and mailing address — provides a new opportunity to perform an action on that contact directly from each contact's details screen.

To get to a contact's details screen, just touch the contact's row in the Contacts tab. Let's look at an example in Figure 8-6.

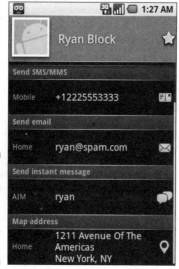

Figure 8-6: A contact details screen filled with plenty of info.

Here, we see just a portion of the contact's full information — but from this snippet alone, we can do lots of things. Each section, denoted with a gray header, indicates a type of activity. Below each header are the details that correspond to that activity.

The simplest activity is Dial number. Touching a number in this section dials the number. But the fun's just beginning! When you add a phone number to a contact, that number also becomes available for the Send SMS/MMS activity, so choosing the same phone number in this section opens the Messaging application and fills out the To box with this number.

What else can I do? If I entered an e-mail address, I can start composing a new e-mail addressed to this contact simply by touching the e-mail address itself. In this case, both the Email and Google mail (Gmail) applications can service my request to send a new e-mail, so I'm given the option of choosing which to use, as shown in Figure 8-7. If I always want to use one or the other, I can select the Use by default for this action check box.

Figure 8-7:
What appli-
cation do
you want
to use for
sending
an e-mail
to this
contact?

But it doesn't stop there. I also see that for the Send instant message activity, this contact is available on AIM as ryan. Touching this detail, therefore, would open the IM application so I could send an instant message via AIM to screen name ryan (more on instant messaging in Chapter 12).

A street address has also been entered, so it appears under the Map address activity. Touching the address opens the Maps application and pinpoints the address for me — a lifesaver if I'm unfamiliar with the neighborhood!

Editing and Deleting Contacts

You can edit a contact in two ways. If you're currently in the Contacts tab — not looking at a specific contact's details — do the following:

1. **Locate the contact you want to edit in the list.**

2. **Touch and hold on the contact's row until the pop-up menu appears.**

3. **To delete the contact, touch Delete contact. To edit the contact, touch Edit contact.**

If, instead, you're already looking at a contact's details:

1. **Push the Menu button.**

2. **To delete the contact, touch Delete contact. To edit the contact, touch Edit contact.**

In either case, touching Delete contact displays a pop-up message asking for confirmation that you really want to get rid of the contact forever. From this message, you can always touch Cancel to back out.

If you're editing the contact, you are brought to the same screen you worked with when you created the contact — familiar territory. When you've finished making changes, simply touch the Save icon at the bottom of the screen or push the Menu button and touch the Save item.

Contacts Settings Screen

The configurable settings for the Contacts tab are surprisingly simple — in fact, you have only two! To get to the Settings screen (see Figure 8-8), push the Menu button while looking at the Contacts tab and then touch the Settings menu item.

Let's take a look at your options here:

- **Sync groups:** If you have stored in Google certain groups of contacts that you know you'll never need on the phone, you can save storage space on your G1 and wireless data usage by limiting which groups are synchronized with Google.

  When you touch this item, a menu appears with a list of group names to synchronize. A check mark to the left of a group name means that it's set to be synchronized. By default, all groups are synchronized, which is set by selecting the Sync all contacts item at the very top of the list. When this is selected, you cannot deselect individual groups — they're all selected.

  To limit the groups that are synchronized, first touch Sync all contacts to remove the check mark, which automatically deselects every group in the list. Then, go through and touch the groups to which you want to add a check mark. You can always revert to the default option of synchronizing all groups by touching Sync all contacts again.

- **SIM contacts importer:** As I mention in Chapter 4, the SIM card installed inside your G1 contains a little bit of usable memory for storing SMS messages and contacts. Back in the dark ages of cellphones — before Android, that is — you could copy a small number of contacts (each with a limited amount of information) between cellphones by moving the SIM from one phone to the next. With Android, this isn't necessary because all your contacts are automatically stored with Google.

  If the SIM card in your G1 has been brought over from another phone, though, and has some contacts stored on it, you can see these entries in your contacts but you can't call them or use the data associated with them until you import them. To perform the import, touch the SIM contacts importer menu item, which brings up a list of all contacts on the SIM. Pressing on a contact in this list displays the New contact screen with as much information already filled in as possible. Add whatever additional information you'd like — just as though you were creating any other new contact — and then touch Save at the bottom of the screen to create a new contact on your G1. You can also import all contacts at once by pushing the Menu button and touching Import all, but beware — all the contacts on the SIM will be imported immediately, and you'll have to go back to each imported contact's details screen to edit any additional details.

Chapter 9

Calendar

*I*f you're still using an outdated PDA, pencil and paper, a few well-placed Post-It notes, or some scribbles on the back of your hand to keep track of your busy schedule, that strategy comes to a screeching halt today. Let's be honest, you lead a crazy life — we all do — and you need all the help you can get keeping track of meetings, appointments, dinners with the family, and everything else that keeps you running from one part of town to the other day in and day out. Well, you may have noticed that Google does a fantastic job of managing your personal information and keeping it stored, safe, and secure. Your calendar is no exception.

Just like your contacts, your calendar is stored in the "cloud" — it's automatically synchronized to Google's Internet-based services. The advantage of that seamless, automatic synchronization is threefold:

✔ **You don't need to use your G1 to see your calendar:** Having your schedule on the go is a must, but for many of us, sitting at a desk in front of a computer monitor is a daily reality. Why would you want to use your G1 to check your schedule when your PC is right in front of you? With Android and Google, all you need to do is head over to `calendar.google.com` to see your upcoming goings-on from the soft, warm glow of your desktop display.

✔ **Backup is silent and automatic:** I'm sure you treat your G1 with the care that you'd treat a child — but accidents happen. If your beloved phone breaks or goes missing, your schedule won't be lost with it. (On the downside, you'll no longer have a good excuse to miss that snoozefest of a status report meeting.)

> ✔ **You can share calendars:** One of the most interesting features of Google Calendar is its ability to support many separate calendars on the same display. For example, maybe you have Work, School, and Personal calendars for yourself — but you also have a Rec League Softball calendar that you share with your team. With the G1, any changes you make to shared calendars will be seen by everyone who shares that calendar with you.

One Calendar, Four Ways

More than anything else, you'll be using the Calendar application to glance at your schedule; of course you can make edits and create new events on-the-fly, but job number one is to show you the goods. Just like Google Calendar on your desktop, Calendar on your G1 can slice and dice your schedule a variety of ways.

First, let's open the Calendar. You do this the same way you open virtually any other application on the G1 — touch its icon in the Applications tab or its shortcut on the Home screen if you've created one.

From time to time, you might get an e-mail from Google that contains a link to view a calendar. This can happen, for example, if a contact of yours shares a new calendar with you. These e-mails were designed to be read from your desktop, though, and as of press time these links don't work on your G1. If you click them, you're taken to the Browser, where you see an error message saying that you're using "a browser that isn't currently supported" (see Figure 9-1). For the most part, this isn't a problem — shared calendars are automatically added to your account, so you'll still be able to see any events on the calendar simply by going to the Calendar application as usual.

Once you've opened the Calendar application, you'll see one of several views. (I get to these next.) If you already use Google Calendar on the Google account to which your G1 is associated, you see any events there listed here — they're continuously synchronized. In other words, any changes you make from your desktop are visible here without any extra work on your part, and vice versa.

If you don't use Google Calendar yet, you'll see a clean slate — but don't worry, we'll fix that soon enough!

The very first thing you'll want to be able to do is cycle between views. You can choose from four: Day, Week, Month, and Agenda. From any view in the Calendar, you can zip on over to a different view simply by pushing the Menu button and making a selection (see Figure 9-2); the selections appear as the first three menu items (and naturally, the view you're currently in will not be an option).

Figure 9-1:
Ironically,
the Web-
based
version
of Google
Calendar
isn't work-
ing on the
G1 just yet!

Figure 9-2:
Choosing
a view.

Day view

The Day view, shown in Figure 9-3, is a complete picture of the events in a particular day, from 12 a.m. to 11:59 p.m.

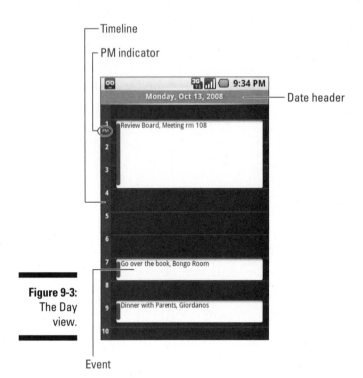

Timeline

PM indicator

Date header

Figure 9-3:
The Day
view.

Event

Let's take a look at how the view is laid out:

- **Date header:** The day of the week and date are always visible as a gray bar directly below the status bar, even if you've scrolled down to the end of the day.

- **Timeline:** Each hour from 12 a.m. to 11 p.m. appears on the timeline to the left of the display. An *am* or *pm* indicator is visible on the first hour at the very top, no matter what hour you're looking at — that way, you always know whether you're looking at morning or evening (a good thing; otherwise I'd be showing up for dinner dates at 7 in the morning).

- **Events:** Your events appear throughout the timeline as white boxes with the name of the event and the location (if you specified it) written within. If multiple events overlap, they'll be narrower so they can all appear across the screen. All-day events, which are events that aren't assigned explicit start and end times, appear as immovable white boxes near the top of the screen between the date header and the topmost hour marker.

Touching an event gives you additional information about it in a box at the bottom of the screen, as shown in Figure 9-4. For this Review Board meeting I have scheduled from 1 p.m. to 4 p.m., for example, the box explicitly spells out the start and end time of the event (handy, since this can be unclear just by looking at the timeline for events that don't start or end on the hour or half hour). Because I've set a reminder for this event, it also shows a bell icon. This extra detail box covers up part of your timeline, so it disappears after a moment to get out of your way.

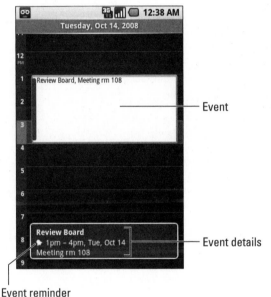

Event

Event details

Figure 9-4:
Extra details
for an event.

Event reminder

To move between hours of the day, swipe up and down or move the trackball up and down; similarly, you can swipe or move the trackball left and right to change days one at a time. You can quickly return to the current date by pushing the Menu button and touching the Today menu item, which also automatically selects the current hour-long time block.

Week view

The Week view, shown in Figure 9-5, is similar to the Day view — in fact, you can think of it as a Day view compressed horizontally to show seven days across. Just like the Day view, you can swipe up and down to move all the way from 12 a.m. at the top to 11 p.m. at the very bottom, and you can swipe left and right to move between weeks.

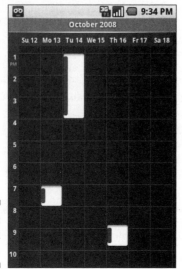

Figure 9-5:
The Week
view.

One difference you'll notice from the Day view is that there are no event details within the white boxes — they're just not wide enough! You can still touch a box to select it, though, which will call up the event details box at the bottom of the screen.

Pushing the Menu button and touching Today returns you to the current week and, like the Day view, places the highlight on the current date and time block.

Month view

Understandably, the Month view (see Figure 9-6) is pretty high level. Let's just say you're not going to be able to figure out exactly what's going on for a particular day from here, but then again, that's not really the idea. The Month view is simply a way for you to get a quick picture of your availability way out into the future.

At the top of the screen is a gray bar that indicates the month and year you're currently viewing. Each day within the month is composed of a block with a number — that day's date — and a vertical white bar to the right. This bar represents the entire day from 12 a.m. at the top to 11:59 p.m. at the bottom, just like a miniaturized Day view. Busy times within the day are denoted with green blocks.

To quickly get to the current month, push the Menu button and touch Today.

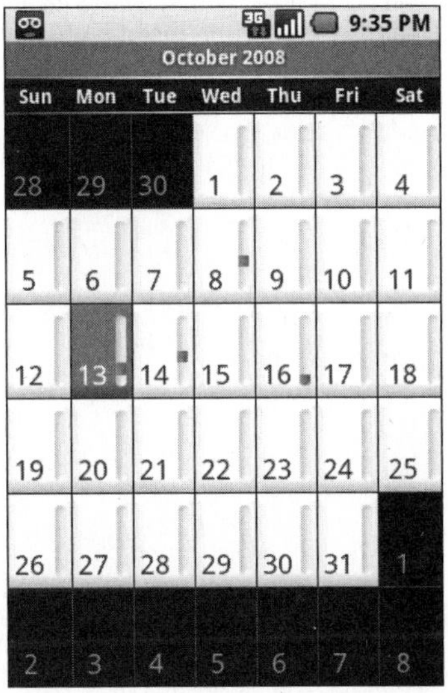

Figure 9-6:
The Month
view.

Even if you configure a particular event to show you as available instead of busy, you'll still see it shown in green in the Month view. In other words, in the Month view, all events are shown in green.

From here, you can swipe up or down to move between months. If you're looking at the current month, the block for today's date is shown in gray instead of the usual white. Want to zoom in and figure out exactly what's going on for a particular day? No problem — simply touching any date will take you to its Day view.

By touching and holding a day in the Month view, you'll be presented with a pop-up menu. From here, you can quickly choose to see the Agenda or Day view or create an event to occur on that day.

Agenda view

The Agenda view, shown in Figure 9-7, is a little different than the other three views because it cuts out all the fat — no empty, uneventful times are shown. It gets straight down to business by showing you only the scheduled events around the current time. This view provides a detailed glance of the meetings you have coming up (and those that have recently passed) without scrolling through hours (or days) of dead space.

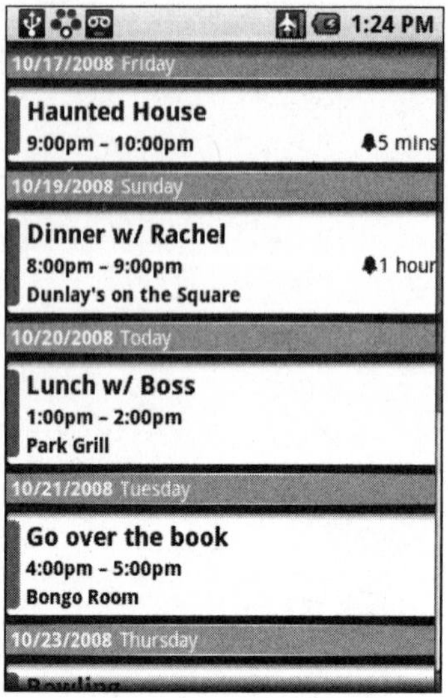

Figure 9-7:
The Agenda
view.

Each day is presented in chronological order as a gray bar with all events scheduled for that day immediately below it. One great thing about the Agenda view is that the events are shown with quite a bit of detail: You get the full name of the event, the start and end times, and the location. If you've set a reminder, you also see a bell icon, and to the right of the bell you see the number of minutes, hours, days, or weeks before the event time that the reminder alarm is set to trigger. (I discuss adding reminders to your events later in the "Creating Events" section.)

From the Agenda view, pushing the Menu button and choosing the Today menu item positions the header for today's date at the very top of the screen, making it easy to see what's coming up next by simply looking at the top and reading down.

Working with multiple calendars

As I mentioned, Google Calendar supports multiple calendars, which makes it a breeze to share events with others, separate your work from your personal life, manage different projects or teams, or employ pretty much any other sort of organizational technique you can dream up. If a contact of yours shares a calendar with you, you'll automatically see it on your G1 by default, and you can also go to `calendar.google.com` from your PC to create new calendars of your own.

You may have noticed in the previous screen shots in this chapter that events have different colored vertical bars (or different shades of gray in this case, since the book is printed in black and white) to the left side of their boxes. This is an indication of what calendar they belong to. From your G1, you can determine what color corresponds to what calendar, and choose which calendars you want to see. (Your view can get cluttered quickly if too many calendars' worth of events are shown at the same time.) To do this:

1. **From any view in the Calendar application, push the Menu button.**

2. **Touch the More menu item.**

   A new submenu appears with additional items not available on the first menu you saw.

3. **Touch My calendars.**

   You see a screen similar to Figure 9-8 that presents a list of all your calendars (and calendars that are shared with you) that are currently synchronized to your G1.

4. **To hide a calendar from view, simply touch it to deselect it.**

Figure 9-8:
Selecting
calendars
to display.

From the My calendars screen, you can also remove calendars altogether so that they are no longer synchronized to the G1, saving space on your device and reducing data traffic. Here's how:

1. **Push the Menu button.**

2. **Touch the Remove calendars menu item.**

   In the window that appears, touch to place a check mark next to any calendar that you want to remove from your G1.

3. **Touch OK.**

Removed calendars still exist in your Google account, and you can view them by going to the Google Calendar Web site from your PC. You can add calendars back for synchronization with your G1 by repeating the preceding steps, touching Add calendars instead of Remove calendars in Step 2.

Creating Events

From any view in the Calendar, you can push the Menu button and touch the New event item to get the ball rolling on the creation of a new event, but you can do it using a couple of other methods too:

- ✔ From the Day or Week view, highlighting an open time (either by touching it or rolling the trackball to it) and selecting it will create a new one-hour long event starting at the time you selected. (You can adjust the time and duration from the New event screen.)

- ✔ From the Month view, hold down on a date until a pop-up menu appears and then choose New event to create a new event scheduled for the selected day.

Any way you initiate the creation of the new event, you'll end up with a screen like Figure 9-9.

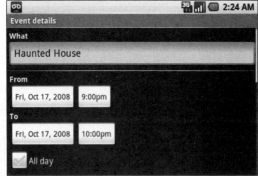

Figure 9-9: Creating a new event.

Let's go over the available fields:

- ✔ **What:** This is the name of the event — for example, "Lunch" or "Shop for some cool For Dummies books." At a minimum, you'll want to fill this out so you know what the event is.

- ✔ **From/To:** You'll see two boxes for both the From and To fields — one each for the date and time. Touch any box to modify it. You can use the + and — icons on the screen to increment the month, date, year, hour, and minute, but you can also touch the actual text in the center and type right over it as you would for any other text box. Use whatever method is easier for you. When you have each of the four boxes set to match the start and end times for this event, touch OK to accept your entry or Cancel to discard it.

✔ **All day:** If the event you're scheduling runs all day, such as vacation, you can add a check mark here. When you do so, the times from the From/To fields above disappear, and you'll be able to select only dates.

✔ **Where:** This is the location where the event is taking place. It is just for your reference and is not a required field.

✔ **Description:** You can add any notes about the event here, like "Wow, this is going to be a boring meeting! Glad I'll have my G1 with me for entertainment."

✔ **Calendar:** If you have multiple calendars, you can choose which calendar this event is assigned to (for example, meetings might go to your Work calendar) by touching this field and selecting an option.

✔ **Reminders:** Before the start of an event, you can schedule one or more reminders to appear on your G1. By default, you'll see one reminder here (unless you've changed the default — more on that when I talk about Settings later in the chapter). You can modify how long before the start of the event the reminder fires by touching the down arrow and choosing a duration from the menu.

If you want to get rid of a reminder, just touch the red X icon next to it on the right side of the screen. To add a reminder, push the Menu button and touch Add reminder.

✔ **Repeats:** Most events you create will likely be one-time affairs, but occasionally you'll set up a meeting, an annual gathering, or something else that repeats on a regular basis. Fortunately, Android can handle these with aplomb. Touch the down arrow on the box and choose a repeat interval, and Android adds the appropriate events automatically.

If you push the Menu button and touch Show extra options, the New event screen adds two additional fields at the bottom of the list:

✔ **Presence:** This lets you select whether other people with free/busy access to your calendar see this event as available or busy time. The default option is Show as busy, but there may be times (such as a meeting you don't plan on attending, for example) when you want to add an event to your calendar while still showing yourself as being available to others who can see your free/busy status.

✔ **Privacy:** This option sets who can see details about this event. Calendars themselves have privacy settings, and the default option — appropriately named Default — just makes the event use its calendar's settings. You can override that by choosing the Private or Public option.

After you've finished entering all the information for this event, just touch Save to commit it to your calendar or touch Discard changes to cancel it.

Working with Events

You can touch any event in any view to be brought to the View event screen (see Figure 9-10), which lists everything there is to know about the event: name, time, location, name of the calendar on which it resides, and any reminders associated with it.

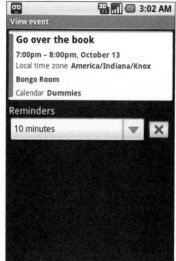

Figure 9-10:
The view
event
screen.

From here you can add and remove reminders in the same way you do when you're adding or editing an event. Push the Menu button and touch Add reminder to add a reminder, or touch the X next to the reminder to remove it.

Invitations

If this event was sent to you by another contact as an invitation, you see a field with the Attending? label. The default is (No response), but this can be changed to Yes, Maybe, or No to notify the sender of your intentions.

Unfortunately, you can't currently use your G1 to invite others to attend events in your calender. To do that, you need to use the Google Calendar Web site from your PC.

Editing and deleting events

To edit or delete an existing event:

1. **Find the event you want to edit in the Day or Week view.**

2. **Touch and hold down on the event until a pop-up menu appears.**

3. **Touch Edit event or Delete event.**

   Choosing Edit event returns you to the same screen you used when creating the event — from here, you can modify fields as necessary and then touch Save at the bottom of the screen. Touching Discard changes keeps the event but throws away any changes you've just made. Delete removes the event altogether.

You may not have permission to change some of the calendars you see on your screen that other people have shared with you. For events on these calendars, you will be able to only view the event from the pop-up menu, not edit or delete it.

Calendar Settings Screen

To show the Settings screen for the Calendar app, push the Menu button from any view, touch the More menu item, and choose Settings. This will bring you to a display like the one in Figure 9-11.

Here's what you can configure from the Calendar's Settings screen:

- **Hide declined events:** If this is selected, events you are invited to that you indicate you will not be attending are not shown on your calendar.

- **Set alerts & notifications:** Touching this lets you choose from three styles of reminders: Alert, Status bar notification, and None. Alert is the most in-your-face option, covering your entire screen with a message detailing the event, as shown in Figure 9-12. Status bar notification is your typical notification — just like you'd get for text messages, e-mails, or other features — and None just quietly ignores your reminders.

- **Select ringtone:** This selects the sound to accompany reminders.

- **Vibrate:** If this check box is selected, your phone gives you a little jolt when your reminders go off.

- **Set default reminder:** This setting determines the default reminder duration used when you create a new event. If you choose Never, you need to add a reminder manually to each event you create for which you want a reminder.

Figure 9-11:
The
Calendar
application's
Settings
screen.

Figure 9-12:
An event
alert.

Part III

Making the G1
Part of Your Life

The 5th Wave

By Rich Tennant

"I find it so obnoxious when people use their cell phone in public that I'm getting on the Web right now to blog about it."

In this part . . .

There's a whole lot of Google stuffed into this phone, but the G1 is so much more than that. In Part III, you're going to find out how to surf the Web, talk to friends through services such as AIM, customize settings, and bring out the G1's lighter side with pictures, YouTube videos, and music.

Chapter 10

Browsing the Web

*U*nless you've already cut your teeth on a device like the iPhone, odds are you're not accustomed to being able to browse the World Wide Web — yes, the real one, not a watered-down one — right on your handset. Although virtually all cellphones sold today offer Web browsing capability in one capacity or another, few can effectively surf sites with a level of sophistication that nearly matches that of your PC. With these less-capable phones, sites you want to browse don't load, don't look right, or are presented as a mobile version that doesn't offer all of the regular site's functionality. Who wants that?

Fortunately, the G1's Browser application (see Figure 10-1) is one of the best phone-based Web browsers the world has seen. It can do just about everything your desktop browser can, and thanks to the G1's support for Wi-Fi and 3G data networks, it can do it all downright speedily. In this chapter, I get into everything the Browser has (and doesn't have) to offer, explain the differences between browsing on your G1 and browsing on your desktop, and show you some cool tricks along the way that'll darned near make you forget your desktop's browser ever existed!

Figure 10-1:
The
Browser.

Opening the Browser

If you want to choose a bookmark or browse directly to a particular Web address by typing its URL, you can do so simply by touching the Browser icon in the Applications tab (or on your Home screen, if you have an icon for it there). If you want to perform a Web search, though, you don't even have to start the Browser first — you can enter your search from the Google search widget (if you have it on your Home screen) or by pushing the Menu button while at your Home screen and touching Search, as I mention in Chapter 5.

WebKit: The engine that makes it all happen

Although many Web browsers are available for PCs and mobile devices, only a handful of popular *browsing engines* power those browsers. Each engine is a jumble of computer code that decides how to interpret HTML and other Internet languages and render it on the screen so that you can see it. Because each engine behaves slightly differently, Web site designers need to make sure that their sites are fully compatible with all of the most popular engines; otherwise, they risk losing readers!

The four most popular engines in the world for Web browsers are Trident, Gecko, WebKit, and Presto. Trident is used by Microsoft Internet Explorer, Gecko is used by Firefox, and Presto is used by Opera. WebKit, however, is used by a variety of moderately popular browsers, including Apple Safari and Google Chrome.

WebKit is used also by a number of browsers designed for cellphones because it's renowned for its compactness and speed, attributes that are especially important on phones, where memory consumption and processor power are at a premium. Mobile Safari (used by the Apple iPhone) and the S60 browser (used by Nokia's smartphones) use WebKit, and it's no coincidence that they're both highly regarded as fast, stable, and powerful mobile browsers. Android's Browser application also uses WebKit, which is a good thing — anyone who's used Mobile Safari on an iPhone can tell you that!

Additionally, you can touch (or select with the trackball) any Web address in another application (such as Gmail) to open the Browser and go straight to that address — no need to memorize the address or copy it and open the Browser the old-fashioned way.

Going to or Searching for a Web Address

Once you're in the Browser application, the easiest way to go to a Web address is by simply starting to type it (you'll need the keyboard open to do this). As long as you don't currently have a text box selected in a Web page that allows you to type into it, the Browser will detect that you're trying to enter an address and automatically call up an address bar to allow you to enter your URL.

As you type, the G1 will search your browsing history and look for addresses that match what you're typing. If it finds any matches, they are displayed as a list below the text box (see Figure 10-2). If you see what you're trying to browse to, simply touch it in the list (or select it with the trackball) and the Browser will browse to it, which saves you a precious few keystrokes.

Address and search bar

Figure 10-2:
The
Browser's
address and
search bar.

Browsing history

As I mention in Chapter 5, the Browser's even smarter than that, too! It turns out that the address bar is also a search bar — you can use it to type a Google search instead of a Web address.

Once you've finished entering your address or search terms, simply press Enter on the keyboard or touch Go to send the Browser on its way. The gray bar directly below the status bar indicates the Web address of the site being opened and the name of the site (if available); a rotating circle to the right tells you that the Browser's hard at work downloading data. The gray bar progressively changes to yellow until the site — all its textual and image content — has finished loading, at which point the bar fades back to gray.

You can type Web addresses and perform searches using another method if you currently have a text box on a Web page highlighted. (This method works even if you don't have a text box highlighted, but it's especially handy if you do.) The Browser doesn't recognize that you're trying to type a new address because the characters you type go into the Web page's text box. To type a Web address this way:

1. **Push the Menu button.**

2. **Touch Go to URL.**

   A box appears at the top of the screen where you can enter a Web address.

To enter a Google search, the technique is slightly different:

1. **Push the Menu button.**

2. **Touch Search.**

   A Google search box appears at the top of the screen. This differs from the address box you've seen using other methods in that you can enter only searches.

Support (or the lack thereof) for Adobe Flash

As you zip around the Web on your desktop browser, you'll notice that a good percentage of the sites you visit have animated or dynamic content — objects that move, glow, or change based on actions you perform. This content is often powered by a technology called Flash, made by Adobe.

Because Flash requires lots of horsepower to run smoothly and look good, mobile phones with Web browsers frequently don't include it for fear that you'll run out of memory or overextend your processor so that the entire phone becomes very slow to use. Nokia's S60-powered devices usually include a special version of Flash called Flash Lite — but they're the exception to the rule.

At the time of publication, Android doesn't support Flash in the Browser application. However, Adobe is excited about making Flash available to as many devices as it possibly can, so I believe that it will eventually be available as a download. In the meantime, you may see some empty boxes around the Internet (and the occasional page you simply cannot use at all) because Flash support is required.

Navigating the Web

Just like the Web browser on your PC, Android's Browser has commands that allow you to navigate back and forth through Web pages you've recently viewed.

Moving backward and forward

To move back to the page you were at immediately before the current page, simply push the Back button. Alternatively, you can do the following:

1. **Push the Menu button.**

2. **On the menu that appears at the bottom of the screen, touch More.**

3. **Touch Back.**

Believe it or not, there's yet a third way to go to the previous page: Hold down the Menu key on the keyboard and press J. In other words, there's no shortage of ways to retrace your steps!

If you've moved back to a previous page and you'd like to move forward again, there's no Forward button you can push — but not to fret:

1. **Push the Menu button.**

2. **Touch the More menu item.**

3. **Touch Forward.**

Like Back, there's a key combination that can be used as well — hold down Menu on the keyboard and press K (conveniently, this is adjacent to J, so the key combinations for Back and Forward are right next to each other).

Selecting links

To follow a link on a page — that is, to navigate to a page represented by a link — simply touch the link. Alternatively, you can move the trackball around, which will cause a yellow box to appear on the screen (see Figure 10-3). As you move the trackball, the box will highlight any links it encounters. When the link you want to select is highlighted, press down on the trackball.

Figure 10-3:
Highlighting
a link in the
Browser.

Refreshing the current page

Sometimes you're looking at a page that is updated frequently — say, the score in that all-important Cubs-Cards game. To refresh the current page (that is, retrieve the latest version of the page from its source), do the following:

1. **While looking at the Web page that you want to refresh in the Browser, push the Menu button.**

2. **Touch the Refresh menu item.**

History

Remember that great Mediterranean recipe site you found yesterday? If only there were an easy way to go back and find it. Oh, that's right, there is! Just like your desktop browser, the Browser app on Android puts your browsing history just a few clicks (or, in this case, touches) away.

To get to your history:

1. **Push the Menu button.**

2. **Touch the More menu item.**

3. **Touch History.**

The Recently-visited pages screen is called up (see Figure 10-4), showing every page you've visited recently in chronological order, with the most recent at the top.

Each day is broken up as its own section, titled by a gray bar. At the top we see Today — assuming you've visited any pages today — and further down you see a gray bar for Yesterday, and so on.

Recently-visited pages

Yesterday

http://www.blogcdn.com/www.engadget.co...
http://www.blogcdn.com/www.engadget.com/media/2008/10/10-18-...

Engadget
http://www.engadget.com/

Google
http://www.google.com/m

Apple OS X 10.5 "Leopard" officially debuts -

Figure 10-4:
The
Recently-
visited
pages
screen, also
known as
History.

Touch an item in the list (or select it with the trackball), and the Browser will navigate to it. Touch and hold down an item here instead, and you see a pop-up menu of commands. You'll find a surprising number of things you can do with these addresses. Let's walk through your choices:

- **Open:** The first option simply directs the browser to navigate to this link. It's the same as touching the item without calling up the menu.

- **Open in new window:** If you already have the Browser looking at a page that you're not finished with, you can open this link in a new window — just like tabs on your desktop browser. I discuss multiple browsing windows in detail later in this chapter, in the "Working with Multiple Windows" section.

- **Bookmark link:** If you want to add this link to your bookmarks to make it easily accessible to you on a long-term basis, this option is the way to do it. I get into bookmarks in more depth in the next section.

- **Share link:** This option is a cool way to quickly and easily share a Web page with friends, family, or coworkers. By touching this selection, Gmail opens and creates a new e-mail with the selected Web address filled out for you in the e-mail's body. All you need to do is fill out the To field and, optionally, the Subject field and anything else you'd like to say in the e-mail's body.

- **Copy link URL:** This command copies the Web address of the selected item so that you can paste it somewhere else. The copied address can go anywhere in Android where you see a text field — a calendar entry, an e-mail, a contact's details — the list goes on.

- **Remove from history:** If you want to remove just this one Web page from your browsing history, select this item. The remainder of your history will be preserved.

If you want to clear your entire browsing history, push the Menu button while you are viewing the Recently-visited pages screen and then touch the Clear history menu item.

Bookmarks

If you're familiar with virtually any browser on your PC, odds are you're familiar with bookmarks (known as favorites in some browsers). The concept is simple and works the same on Android. *Bookmarks* are simply a collection of Web addresses that you browse frequently and want quick access to without having to remember and type the address each and every time.

Adding a bookmark

Bookmarking a page that you're currently looking at is a breeze:

1. **Push the Menu button.**

2. **Touch the More menu item.**

3. **In the list that appears, scroll down (either with your finger or the trackball) until you find the Bookmark page item.**

4. **Select Bookmark page.**

   A new window appears, titled Bookmark link (see Figure 10-5). The Name and Location fields are filled out for you with the name of the page you're attempting to bookmark (if available) and its Web address, respectively, but you can modify these by selecting the field and typing. You might modify the Name, for example, to something more memorable or meaningful to you.

5. **After the Name and Location fields are to your liking, touch OK to add the page to your bookmarks.**

Figure 10-5:
The
Bookmark
link window.

The bookmarks screen

So where did your newly created bookmark go? Bookmarks go to a dedicated Bookmarks screen, which can be accessed while looking at any Web page:

1. **Push the Menu button.**

2. **Touch the Bookmarks menu item.**

   A screen similar to the Recently-visited pages screen appears (see Figure 10-6), except the Bookmarks screen shows only pages you have specifically bookmarked.

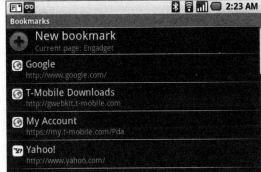

Figure 10-6:
The
Bookmarks
screen.

The G1 comes with several bookmarks to popular Web sites such as Wikipedia, CNN, Amazon, and the Weather Channel, along with entries for Google and T-Mobile's Downloads and My Account pages. (Of course, you can delete any of these.)

At the top of the Bookmarks screen is another way to add a bookmark. Touching the New bookmark item bookmarks the page you were just looking at before coming to the Bookmarks screen. (The Current page line below New bookmark reminds you of which page that was.) Alternatively, you can push the Menu button from this screen and touch Bookmark last-viewed page. In either case, you are presented with the Bookmark link window (refer to Figure 10-5), where you have the opportunity to edit how the name and Web address of your bookmark will appear in the Bookmarks screen.

Using, editing, and deleting bookmarks

To select a bookmark and navigate to it using the Browser, simply touch it or select it using the trackball. Alternatively, you can touch and hold on a bookmark, which calls up a pop-up menu; then touch Open or Open in new window.

The pop-up menu I just mentioned lists several other things you can do with your bookmarks besides open them:

✔ **Edit bookmark:** Selecting this option brings up the same window you see when adding a new bookmark, except the window is titled Edit bookmark instead of Bookmark link. Here, you can adjust the name of the bookmark or the Web address to which it points. Touch OK to commit your changes or Cancel to discard them.

✔ **Share link:** This option works the same way as it does in the Recently-viewed pages screen. Selecting Share link adds the bookmark to the body of a new e-mail and opens the e-mail for editing.

✔ **Copy link URL:** This item copies your bookmark's Web address so you can paste it elsewhere in the G1.

✔ **Delete bookmark:** Select this option to delete the bookmark from your Bookmarks screen. You receive a confirmation window asking you to make sure you want to delete the bookmark; touch OK to go ahead with the deletion.

Scrolling and Zooming Web Pages

Although the Browser application is fantastic at showing the full content of a Web page exactly as you'd see it on your PC, there's a problem with that: Web pages are big, and your G1's screen is small! The good news is that Google has cooked up a number of creative ways to get around this limitation, and once you become familiar with these tricks, you'll be a pro at navigating huge pages on that tiny (well, relatively tiny) display in no time.

Scrolling

To scroll around a Web page that's too big to display on the G1's screen (which will be many — if not most — of them), simply touch and hold your finger to the display and move it around. Think of it like placing a hockey puck on an ice rink and moving it around with your finger — the Web page moves in the same way.

By stopping your finger's movement before lifting it off the screen, the Web page will remain where you left it. However, if you *flick* your finger across the screen — meaning your finger is still in motion at the time that it leaves the screen's surface — the Web page continues to move for a while before it slows to a halt. Flicking, a convenient way of scrolling across Web pages quickly, is a surprisingly natural motion, and you'll become used to it in no time.

This isn't the only way to scroll, however; you can also use the trackball if you prefer. As you move the trackball around, the yellow highlight box I mentioned earlier in the chapter appears and highlights any links it comes across. At any time, you can press the trackball to follow a link.

Zooming

Sometimes, scrolling isn't enough to help you get the big picture of what's going on with a particular Web page — you need to see more, and you need to see it all at once. How do you make that happen? In this section I go through your options for zooming pages in and out.

You may have noticed that when you scrolled around the page with your finger, some icons appeared at the bottom of the screen (see Figure 10-7). These are the Browser's zooming controls. All told, the Browser offers a whopping *14* levels of zoom for the pages you view, and you have access to all of those levels using these controls. (You can call up these controls without scrolling the screen by pushing Menu, touching More, and touching Zoom from the menu that appears.)

Figure 10-7:
The Browser's on-screen zooming controls.

Zoom in   Zoom out   Four-way arrow

To zoom in one level, simply touch the magnifying glass icon with the plus sign (the left icon). To zoom out one level, touch the icon to the right of that — the magnifying glass with the minus sign. Easy, right? At every level of zoom, your scrolling controls still work the same — just move your finger or the trackball to navigate.

Although it's difficult (well, nearly impossible) to read normal-sized text at the outermost zoom level, this level of zoom is a good way to see very large images that you encounter on your adventures across the Internet.

The Browser's zooming capabilities get cooler, though — way cooler! By touching the four-way arrow in the lower right of the screen, the Browser zooms all the way out and shows a magnifying box in the middle of the screen (see Figure 10-8). By touching this box and dragging it around the screen, you can specify what part of the page you want to get a closer look at; when you let go, you return to the previous zoom level. Think of the zoomed-out view as a quick way to get a bird's-eye view of the entire Web page to help you decide what part you want to see up close on your G1's screen. See, browsing big Web sites on a small screen isn't so hard, is it?

Figure 10-8:
The
Browser's
magnifying
box.

While viewing a Web page, simply press the trackball twice quickly — a double press — and you'll get the same bird's-eye view I previously described. The difference here is that you can use the trackball to control the magnifying box, which makes it way easier to position precisely if you're having a hard time controlling the box with your finger.

Working with Multiple Windows

If you're using a modern browser on your PC — Internet Explorer 7, Firefox, or Safari, for example — you're probably acquainted with the concept of *tabbed browsing*. Put simply, tabbed browsing places each page in a separate tab across the top of the window.

Android has adapted the tabbed browsing concept from PCs to the Browser. In this case, you're not working with tabs in the traditional sense of the word — the screen's just not big enough to devote space to them — but you still get all the functionality that goes along with them. The Browser calls these loaded pages *windows* instead.

Opening a link in a new window

If you're browsing a page and want to follow a link without navigating away from the current page, you can open the link in a new window. To do this:

1. **Touch and hold on the link (or select it with the trackball and hold the trackball down).**

   After a moment, you see a pop-up menu of commands you can perform with the link.

2. **Touch Open in new window.**

   Something amazing happens — the page you're currently looking at zooms out and becomes a square in a grid. (This grid is a view of all the Browser windows you currently have open, which we take a look at next.) Then the new link you selected zooms into view. Magically, you're looking at the new page, and the page that you just navigated from is still in the background so you can return to it at a moment's notice.

Switching between multiple windows

Say that you're finished looking at this new link you've opened in a new window, and you're ready to return to the page you were looking at before. Let's take a look at how to do this:

1. **Push the Menu button.**

2. **Touch the Window menu item.**

   You go to the Current windows screen (see Figure 10-9), which is your home base for managing all the windows that are currently open in the browser. Each window gets its own box in a grid; the box shows a small preview of the page along with the page's title at the bottom. If you have more windows open than will fit on the screen, you can scroll the screen to see all available windows.

3. **Touch the box representing the page you want to view, or highlight it with the trackball and push down.**

   The page zooms into view, and the Current windows screen disappears. To return to the screen again, simply repeat these steps.

Figure 10-9:
The Current
windows
screen.

Opening a new blank window

If you want to keep your current Web page open and use a new window for
viewing a Web address or Google search, that's easy, too:

1. **Push the Menu button.**

2. **Touch the Window menu item.**

3. **Touch the box that shows a page with a plus sign on it and a New
   window label at its bottom, or select it with the trackball and push
   down.**

   A new, fresh window appears and loads your home page. From here,
   you can navigate to a Web address or enter a search term as you
   would normally.

While in the Current windows screen, another way to open a new Browser
window is to push the Menu button and touch the New window menu item.

Closing windows

When you're finished using a window, it's a good idea to close it. Doing so
not only unclutters your Current windows screen but also keeps memory
usage to a minimum. (The G1 is a powerful phone, but memory is always at a
premium!)

To close a window, return to the Current windows screen and touch the X
icon in the lower-right corner of the box representing the page you want to
close. You are prompted to confirm that you want to go ahead with the clo-
sure (see Figure 10-10). Touch OK to go ahead or Cancel to keep the window
open.

Figure 10-10:
Closing a
Browser
window.

Saving Files from the Browser

The problem with the Web is that it's transient — what's here today might be gone tomorrow. You'd probably like to hang on to lots of stuff out there on the Web, and the Browser makes that possible for common things such as pictures and sounds.

Pictures

On occasion, you may encounter a picture on the Internet that you'd like to save to your G1's memory to use as wallpaper or an icon for a contact or to send in an e-mail. With the Browser, it's no sweat:

1. **While looking at the page in the Browser containing the image you want to save, scroll to the image so that it is in view.**

2. **Touch and hold on the image.**

   You see a pop-up menu of commands.

3. **Scroll down in the menu and touch the Save image item.**

   A Download history screen appears, as shown in Figure 10-11, displaying everything you've downloaded or are currently downloading.

What's neat is that you don't need to remain on this screen while you wait for your download to complete — you can push the Back button to return to the Browser (or go to any other application on your phone) and the download will continue in the background. An icon depicting an animated down arrow appears in the status bar while the download is in progress; when it's complete, the arrow stops animating.

By pulling down the Notifications screen and selecting a completed download (see Figure 10-12), you'll be taken straight to the necessary program for viewing or playing the type of file you downloaded, which in this case is the Picture Viewer for viewing the image you just downloaded.

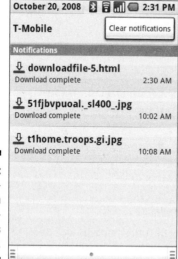

Figure 10-12:
You're noti-
fied when
your down-
load has
finished.

Music, Web pages, and everything else

As I alluded to in the preceding section, your downloads aren't limited to pictures. You can download music files such as MP3s and WAVs, and you can even download Web pages to HTML files on your G1 and view them later. If you want to download something other than an image, you access it via a link instead of an image, but the process is essentially the same.

Say you want to download a music file that is accessed from a link on a Web page and then save the file to your G1 so you can listen to it later. Here's what you do:

1. **In the Browser, locate the link to the music file on the page.**

2. **Touch and hold on the link.**

   After a moment, you get a pop-up menu.

3. **Scroll down in the list and select Save link.**

   As before, you're taken to the Download history screen, where the status of your download (among any other downloads you may have in process) is shown.

If you had chosen the Save link menu item for a link that goes to a Web page instead of a music file, the page's HTML would be downloaded for you to view.

If you attempt to download a file type that isn't supported by Android (such as a Windows program or a PDF), the Download history screen and the Notifications screen both let you know with a message (see Figure 10-13). You don't have to perform any other action — the G1 is smart enough to not download the file, and no space is taken up in your phone's precious memory.

Figure 10-13:
Oops! This
file type
can't be
downloaded
to the G1.

Working with the Download history screen

The Download history screen, which works a lot like your Recently-viewed pages screen, shows you what you've downloaded (instead of what you've viewed) and can be cleared.

First, we need to open the Download history screen. You don't need to start a download to get to it; you can view the screen through the Browser's menu:

1. **While looking at a Web page in the Browser, push the Menu button.**

2. **Touch the More menu item.**

3. **In the menu list that appears, scroll down and select Downloads.**

To clear your entire download history from the Download history screen:

1. **Push the Menu button.**

2. **Touch Clear list.**

   You are warned that all items in the list will be cleared and also removed from the Browser's cache, meaning that you won't be able to open downloaded items from here anymore. And unless the items have been saved to another program such as Pictures or Music, they'll be gone, period. This action affects downloaded Web pages and APK (Android package) files, which are applications you can install.

3. **Touch OK to continue clearing the list or Cancel to abort the operation and leave the list intact.**

You can also work with individual items in the Download history screen. Touching and holding on any item in the list calls up a pop-up menu of operations that you can perform. You see different options depending on the item.

If an item that you select is still in the process of downloading, touching Cancel download stops it in its tracks. If the item has already finished downloading, you see two options:

✔ **Open:** This option simply opens the item. The way the item is opened depends on the type of item. For example, Picture Viewer opens an image and Music opens a sound clip.

✔ **Clear from list:** If you just want to remove a single item from your download history without clearing the entire list, this is the option to use. Selecting this menu item immediately clears the selected item without prompting you to confirm, so make sure you mean it!

Changing the Screen Orientation

Opening the keyboard automatically flips the screen from a portrait to a landscape orientation for you. But because most Web pages are designed to be read on a screen that's wider than it is tall, there may be times when you want to put the G1's screen in a landscape orientation even when the keyboard is closed. To do this:

1. **Push the Menu button.**

2. **Touch More.**

3. **Scroll down in the menu that appears and select Flip orientation.**

To change back to portrait orientation, simply repeat these three steps. Note that the Flip orientation menu item is grayed out — that is, it's not selectable — when the keyboard is open. And really, wouldn't it be a little weird to use the keyboard when the screen's facing the wrong way, anyway?

Browser Settings Screen

With so many similarities between Android's Browser and your desktop equivalent, you might suspect that similarities exist in their settings, too — and you'd be right. Let's take a close look at everything that's available in this screen. First, let's open the Settings screen. Unlike some other applications that make their Settings screen accessible from the first menu that appears when you push Menu, the Browser buries its Settings screen just a little deeper:

1. **Push the Menu button.**

2. **Touch More.**

3. **Scroll to the bottom of the menu that appears and select Settings.**

   The Settings screen appears, as shown in Figure 10-14.

So what sort of settings, gadgets, and doohickeys does Android make available to us? What can we tweak? Here's what we have:

✔ **Text size:** This option adjusts the default text size for all Web pages. A smaller size means more text can fit on the screen, but the tradeoff, of course, is that the text becomes less readable. Touching this item will bring up a menu of available sizes; simply touch the size you want to use.

Figure 10-14:
The
Browser's
Settings
screen.

✔ **Block pop-up windows:** Like desktop browsers, the Android Browser has built-in protection to prevent the display of many pop-up windows (frequently, these are used for advertising and can be annoying, to say the least). By default, pop-up blocking is enabled, but you can disable it by touching the check box to deselect it. If you decide to disable this item, pop-up windows appear as new windows in the Browser.

✔ **Load images:** By default, images on Web pages are loaded. (They're the best part, after all!) There may be times, however, when you are on a slow data connection or are trying to conserve how much data you receive, especially if you're roaming on another carrier's network and are being billed by the kilobyte. Deselecting this box will cause only text to be loaded.

✔ **Auto-fit pages:** When this option is selected, the Browser tries to change the layout of loaded Web pages to better fit on your screen. Depending on what you're trying to view, this option can make pages difficult or impossible to read, so if you're having problems, try deselecting this option.

✔ **Enable JavaScript:** Many Web sites use the JavaScript programming language to do cool things that wouldn't be possible with HTML alone. Enabling JavaScript (the default option) makes the Browser run slightly slower, but because so many modern sites rely on JavaScript to look and work correctly, there's little reason to disable it.

✔ **Open in background:** If this option is selected, links that you open in new windows open "silently" — that is, the new window you've opened won't be immediately shown to you. You'll continue to see the same page from which you opened the link, and you'll have to go to the

Current windows screen to open the new one. Choosing this option is a matter of personal taste. Some people prefer to browse and navigate this way, while others hate it. (You say to-MAY-to, I say to-MAH-to.)

✔ **Set home page:** Touching this item brings up a window that allows you to type a Web address to a site that is shown automatically whenever a new Browser window is shown. The default option, Google's mobile page, is a good one, but feel free to change it!

✔ **Clear cache:** To help reduce the amount of time it takes to access Web pages that you've been to before, the Browser automatically *caches* some items such as images and HTML, meaning it saves them into memory. That way, the next time they're needed, the Browser can load them straight from memory instead of requesting and downloading them again. The size and contents of the cache are managed automatically, but you can manually clear the cache by selecting this option and touching OK in the confirmation window that appears.

✔ **Clear history:** Instead of clearing your history of viewed pages from the Recently-viewed pages screen that we looked at earlier in the chapter, you can do it right from here by selecting this item.

✔ **Accept cookies:** To help identify who you are, Web sites frequently use *cookies*, which are tiny nuggets of data stored in your Web browser. These are actually handy and common; for example, if you log in to a site with which you have an account and want it to remember your login information, the site will do this using a cookie. If you want to block cookies from being placed in the Browser, you can deselect this box, but keep in mind that some sites might not function properly.

✔ **Clear all cookie data:** Selecting this menu item and touching OK in the confirmation window that appears removes all cookies that are currently stored by the Browser. Keep in mind that any Web sites you've logged into will lose your login information, and you'll have to log in again.

✔ **Remember form data:** As you type into text boxes on Web pages, this option will remember what you've typed and offer it as a suggestion the next time you encounter the same text boxes in the future. This feature is a nice little time-saver that cuts a few keystrokes out of your day.

✔ **Clear form data:** If you want to reset the suggestions that the Browser has accumulated for text boxes that you've filled out on Web pages, select this item and touch OK to confirm.

✔ **Remember passwords:** This option is essentially the same as Remember form data, but the Browser treats username and password information for logins to Web sites separately (and rightfully so — it's a security concern). If your G1 might be shared with people whom you'd rather not have access to the sites you've logged into — your bank account, for example — deselect this box.

✔ **Clear passwords:** This option clears any usernames and passwords that the Browser has remembered for sites. If you intend to lend your phone to someone for any length of time, you can select this to make sure the person won't be able to log in to any of your secure accounts (even if Remember passwords is selected).

✔ **Show security warnings:** Web sites that encrypt your information to ensure the security of your data (credit card and login information, for example) must comply with certain rules to make sure that the data is truly safe as it travels across the Internet. When this option is selected, the Browser can warn you if it detects any problems with compliance with these rules on a particular site.

✔ **Enable Gears:** Google Gears is a technology developed by (who else?) Google that allows both Google and non-Google Web sites to do some cool stuff such as store information offline so that it can be accessed without even connecting to the Internet. Not many sites use Gears, but several important ones do, such as Google Reader and Google Docs. I recommend leaving this option selected unless you have a specific reason to disable it.

✔ **Gears settings:** Selecting this item brings up a window showing you what Web sites you've accessed that have attempted to use Google Gears. You can specifically prevent certain sites from accessing Gears here if you so choose; you might block a site, for example, because certain Gears features allows sites to determine your physical location.

✔ **Reset to default:** If you want to give the Browser that new-browser smell, this is the way to make it happen. Selecting this and touching OK resets all settings in this screen to their factory-new default.

There are a *lot* of settings in this screen — more than you'll find in most G1 applications. Of course, you don't need to change them all this second, but you might want to dog-ear this part of the chapter in case you'd like to revisit the settings in the future.

Chapter 11

Working with and Taking Pictures

For a phone, the G1 has an amazingly large, high-resolution display (3.2 inches diagonally and 480 x 320 pixels, for the record) that's capable of displaying bright, vibrant color. It'd be a shame to let that kind of equipment go to waste, wouldn't it?

Well, luckily, you don't have to. Android includes a fairly powerful picture viewing application, aptly named Pictures, which you can use to organize and display the images that are important to you. In fact, you can even perform some basic editing tasks — all without transferring your pictures to a PC.

What's more, the G1 has a 3.2-megapixel autofocus camera that you can use to take some decent shots when you don't have the good ol' point-and-shoot handy. Don't get me wrong: The G1 camera is not about to put your regular-duty camera out of a job, but it'll get you by in a pinch — and it's great for taking candid shots.

The G1 uses the Pictures application to view and manage pictures and the Camera application to take new ones. In this chapter, you look at the capabilities of both applications from the ground up. And what better place to start than with using the built-in camera? You'll be an Android shutterbug in no time — say cheese!

The G1's Camera

Until recently, the kind of photographic power seen in the G1 was unheard of for a cameraphone — and even now, the G1's camera still toward the upper end of the spectrum. The 3.2-megapixel sensor means that the phone can take shots at 2048 x 1536 resolution, and as small as the lens is, it has a mechanical autofocus mechanism — not unlike what you're used to in your regular camera — that allows the camera to take sharp photos at all but the closest distances.

Simply taking a picture is a snap (pun most definitely intended). You can open the Camera application in one of two ways: the traditional way, which is to open it as you would any other application in the Home screen's Applications tab, or the nifty way, which is to simply hold down the G1's Camera button, which is on the side of the phone to the right of the End button (or above the End button if you're holding the G1 in a landscape orientation) until the application opens.

Taking a picture

Once the Camera application is open, you'll immediately see what the camera lens is seeing on your screen; in fact, the entire screen is your viewfinder. To take a picture:

1. **Hold the phone the same way you would a regular camera, with your finger positioned over (but not pressing) the Camera button.**

2. **When you've framed your shot the way you want it, push and hold the Camera button halfway down.**

   You should feel some resistance at the halfway point — stop there. This is the camera's cue to focus the image. After the camera has finished focusing, you'll see a solid green circle in the corner of the screen (see Figure 11-1). Unless your phone is in Silent mode, you'll also hear two quick beeps.

   If the green circle flashes instead of remaining solid (and if no beep is heard when the phone is not in Silent mode), the camera wasn't able to properly focus, so be warned — your picture may not come out sharp. This can happen when you're trying to take a picture of a very dark or low-contrast scene.

3. **Push the camera button completely down.**

   If your phone is not in Silent mode, you hear a shutter sound as the picture is taken.

The camera is focused

Figure 11-1:
The green
circle indi-
cates that
the camera
is focused.

Congratulations! You've just snapped a shot. You're not finished, though. The G1 needs to know what you want to do with the picture you just took. Immediately after you take the picture, you're presented with four options along the bottom of the screen:

✔ **Save:** The first option saves the image to your phone's memory and immediately returns you to the Camera's viewfinder mode so you can take another picture.

✔ **Set as:** If you know *just* the person in your contacts list that this picture should be assigned to, use this option. You have three options that appear if you touch Set as:

- **Contact icon:** Assigns this newly taken picture as the icon for a contact in your contacts list.

- **myFaves icon:** Assigns this picture to one of your myFaves contacts.

- **Wallpaper:** Uses this picture as your Home screen wallpaper.

Regardless of which option you choose, you are taken to a screen where you can crop (trim) the image. The cropping process works in the same way as discussed in Chapter 8 for assigning icons to contacts. Your image appears on the screen with an outlined region within it; only the part of the image within the outline is used. Drag your finger inside the outline to move it, and drag your finger on the outline itself to resize it.

The shape of the outline varies depending on the type of crop that is needed (for example, myFaves icons require a circular image!), but the feature works in the same way.

Once you're satisfied with the shape and size of the crop, touch Save to commit it. If you're not happy with the crop, touch Discard to throw it away.

✔ **Share:** This option lets you immediately send the picture to others using one of several methods:

- **Email:** This option attaches your photo to a new e-mail in the G1's regular, non-Gmail Email application. If you are using Gmail as your sole method of sending and receiving e-mail on the G1, use the next option, Google mail, which will open the Gmail application.

- **Google mail:** This functions the same as the Email option, but instead of using the Email application, Gmail is used. As before, your newly snapped photo is added to a new e-mail as an attachment.

- **Messaging:** This option allows you to send your picture as an MMS message. The Messaging application opens and your image is attached.

  Because the images that the G1 takes exceed the size limit for a multimedia message, the Messaging application offers to automatically resize the image for you so you can send it (see Figure 11-2). If you choose Cancel, the image is not attached.

✔ **Delete:** Let's face it, we've all taken a few less-than-stellar shots in our lives, and this option gives you an easy way to spike the bad ones as quickly as possible. The Delete option immediately deletes the picture you just took (without asking for a confirmation) and returns you to the viewfinder mode — to try again!

Figure 11-2:
Let the
Messaging
app resize
your snap-
shot to
send it in
an MMS
message.

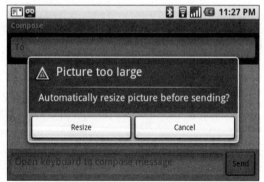

Camera Settings screen

The Camera application's Settings screen can be accessed by pushing the Menu button and touching the Settings item. There are just a couple of options here:

✔ **Store location in pictures:** Touching this check box to select it enables a feature called *geotagging*, which places data about your current location (if your phone can determine it at the time) inside each image file. Many PC programs and Web services (including Google's own Picasa photo management service) can read this data, so you can see exactly where you where when you took a the picture. Kind of cool!

✔ **Prompt after capture:** Enabled by default, after each picture is taken this option shows the Save, Set as, Share, and Delete items described in the preceding section. Deselecting this check box just saves each picture that you take to memory and returns you to the viewfinder to take another picture.

The Pictures Application

Besides the camera, the other piece of the G1's photographic puzzle is its picture viewing capabilities. In this section I take you through everything that the Pictures application can do. Let's start by opening it. Select its icon in the Applications tab, which looks like three Polaroid pictures (naturally!).

If you happen to be in the Camera and you want to hop over to the Pictures application, there's a shortcut — simply push the Menu button and touch the Pictures menu item.

Browsing your pictures

The first thing you'll see when you open Pictures is a thumbnail view of all the pictures on your G1, grouped by album (see Figure 11-3). Each album contains a grid of up to four images. Touching an album (or selecting it with the trackball) takes you to a view of its individual picture thumbnails (see Figure 11-4).

Android automatically uses a few standard picture albums that you should be aware of — you'll see them frequently. The *Camera* album holds any pictures that were taken by your G1's camera. When you're looking at the directory view, you'll notice that this directory gets a special icon — a camera icon — in the lower-right corner of its box. The *download* album contains any images you've downloaded from the Internet using the Browser application. The *sdcard* album contains images copied to the root directory of your G1's memory card from your PC. (I cover connecting your phone to your PC in detail in Chapter 17.)

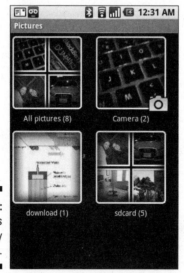

Figure 11-3:
Pictures
grouped by
album.

Figure 11-4:
Individual
picture
thumbnails
within an
album.

By touching and holding on a particular album, you are presented with two
options:

- **Slideshow:** Displays all the images in this album as a full-screen
  slideshow.

- **View:** Views a directory of image thumbnails within the album. This
  option performs the same function as simply touching an album.

Viewing and working with individual pictures

Once you're looking at an image thumbnail view within an album, touching a thumbnail will call up that image as a full-screen view. From here, touching the image will bring up zooming and navigation controls (see Figure 11-5). Similar to zooming in the Browser, touching the plus and minus magnifying glass icons will zoom the image in and out. You can touch and hold on either icon to zoom smoothly and continuously; simply let go to stop zooming.

Previous image   Next image

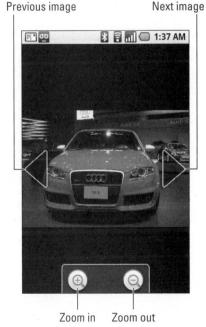

Figure 11-5:
Zooming and navigation controls on an individual image.

Zoom in   Zoom out

When you're zoomed in so that you see only a portion of the full image, you can scroll around the image just as you would a Web page in the Browser: touch and hold on the image and drag your finger around. Touch the left and right arrows on either side of the image to navigate to the previous and next image in the current album, respectively.

By pushing the Menu button while looking at an image, you're presented with a number of options:

✔ **Slideshow:** The first option starts a slideshow of the current album, starting with the current image as the first slide in the show.

✔ **Share:** This option functions the same as the Share option in the Camera application, allowing you to attach the current image to a new e-mail (in either the Email or Gmail application) or to a new MMS message. Note that when sending an MMS, you may be asked to allow the Messaging application to resize a shared image to make it small enough to send.

✔ **Rotate:** Touching this item brings up another menu, allowing you to rotate the image to the left (counterclockwise) or right (clockwise).

When you use the Rotate command, you're actually editing the image — any rotation you apply is immediately saved.

✔ **Flip orientation:** When the keyboard is closed, this item allows you to view the image in a landscape orientation instead of the default portrait orientation.

✔ **Delete:** This option deletes the image from memory. A confirmation window appears. Touch OK to go through with the deletion or Cancel to back out.

By touching More, you are presented with several additional options:

✔ **Crop:** This option calls up the cropping screen (see Figure 11-6), which you may be familiar with from the Set as or Contacts icon command. Like those other screens, you can move the cropping box around by touching and dragging within it. But the difference here is that you're not restricted to a particular shape of box. By dragging on the top or bottom borders of the box, you adjust its height; and by dragging the left or right, you adjust its width. Once you're satisfied with the crop, touch Save. Any part of the image outside the box is discarded and the part inside the box is saved. To cancel the crop, touch Discard.

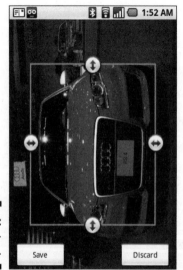

Figure 11-6:
The crop-
ping screen.

✔ **Set as:** This item works the same as the Set as menu item in the Camera application, allowing you to use the current picture as a contact icon, myFaves icon, or your Home screen wallpaper. You'll be asked to crop the image to the correct size before the operation is complete.

✔ **Details:** Selecting this option calls up a window that provides details about the current image — its file name, size, and pixel resolution.

With the exception of Flip orientation, all of these commands are available also by touching and holding on an image thumbnail and using the pop-up menu.

Picture Settings screen

You can access the Pictures application's Settings screen from pretty much anywhere in the application — just push the Menu button and touch the Settings menu item (note that you'll need to first touch More before you're able to select Settings if you're currently looking at an individual image when you push Menu).

Let's have a look at the configurable options that are available here:

✔ **Picture size:** This option determines the size of the image thumbnails you see when you're looking at the screen of all images contained in an album. You have two choices: Large and Small. Unless you're dealing with a *lot* of images on your G1, I recommend using the Large setting because Small is quite small!

✔ **Picture sort**: This option lets you change the order in which images appear within an album — Newest first (that is, at the top left of the screen) or Newest last (that is, at the bottom right).

✔ **Confirm deletions:** If you deselect this option by touching the check box to remove the check mark, you will not be asked for confirmation when you delete an image. Doing so can be dangerous if you're not careful, but if you have a lot of images to delete, disabling this option can save you a bit of time.

✔ **Slideshow interval:** When viewing a slideshow of images, The Slideshow interval option determines the amount of time each image is shown. You can choose between 2, 3, and 4 seconds.

✔ **Slideshow transition:** If you're familiar with PowerPoint or another presentation application, you'll know what this option is all about! This menu option chooses the animated effect that occurs between each image in your slideshow. The default effect is Fade in & out, but you can also choose a left-right slide or an up-down slide, or you can throw caution to the wind and let the G1 decide for you by choosing Random selection.

✔ **Repeat slideshow:** If you select this check box, slideshows loop continuously instead of stopping after they've shown all images within the album.

✔ **Shuffle slides:** By default, images in the slideshow are shown in chronological order, just as they appear in the thumbnail view. By selecting this check box, you can change the behavior so that the slideshow shows images in a random order.

All told, Android's built-in picture management is pretty basic; it's just enough to get you started. The great news is that you'll be able to get some awesome add-on applications for your phone that will let you adjust tint and contrast, add captions, and perform all sorts of cool effects. (Hang tight; I explain how to download those apps in Chapter 18!)

Chapter 12

Instant Messaging

. . . .

In This Chapter

▶ Setting up different instant messaging services

▶ Editing and removing accounts

▶ Adding, removing, and blocking contacts

▶ Chatting with friends, family, and coworkers

▶ Working with the different Settings screens

. . . .

*Y*ou may have gathered by now that Android is talented at keeping you connected. Of course, you have a phone in there — but on top of that, you have all your e-mail accounts and top-notch support for text and multimedia messaging. For staying on the grid, Android is the total package (which could be a good thing or a terrible thing, depending on whether you're trying to get away from work at the moment).

Depending on your T-Mobile plan, you may not have access to unlimited SMS and MMS messages, and if you go over your monthly allotment, you can end up with an expensive bill. Fortunately, the G1 has still other options to keep you in the loop without eating up your text and multimedia messaging allowance. Out of the box, with its built-in software, the G1 supports four major instant messaging (IM) services: AOL Instant Messenger (better known as AIM), Google Talk, Windows Live Messenger, and Yahoo! Messenger. These four are the biggest, most popular instant messaging services available, so the odds are good that you'll be able to chat with most of your PC-based friends — those using an IM service, anyway — from your phone. And with the G1's great keyboard, you'll be able to respond quickly. Heck, they may never even know that you're not in front of your computer unless you decide to spill your secret.

In this chapter, you see how to manage one, two, three, or all four of these services on your G1. You'll be juggling multiple IM conversations and managing contacts with aplomb in no time.

Selecting an Account

First, start up the IM application, which you'll find in the Applications tab.

If you're addicted to instant messaging your friends and family at all hours of the day and night — as so many of us are — you may want to create a shortcut to the IM from your Home screen. You could be accessing it a lot!

The first thing you'll see when you start the IM application is the Chat — Select an account screen, which I'll be referring to as the account selection screen throughout the chapter (see Figure 12-1). You can use this screen to sign in and out of individual services, check your status at a glance (more on that in a bit), and jump into your contact lists for each service.

Let's take a closer look at what we're seeing here. Each of the four services is represented by a row. The icon on the left of each row does more than just look pretty and show you the service's logo. If you look closely, you'll notice a smaller icon to the upper right of the larger one. This smaller icon indicates your status on that particular service (available, busy, away, and so on). These icons match their PC-based counterparts. For example, Google Talk uses a green dot to indicate that you're available to chat, but Yahoo! Messenger uses a smiley face, so you see both styles here on their respective services. If you're not signed in to a service, you see no smaller icon.

Your status (available in this case)

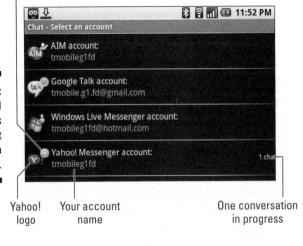

Figure 12-1:
The IM
application's
account
selection
screen.

Yahoo!   Your account   One conversation
logo   name   in progress

To the right of the icon, the service name is displayed along with your account name immediately below it. If you haven't set up an account for a particular service yet, the area where your account name appears is blank. (And rest easy — we'll be rectifying that situation in the next section, where I walk you through setting up accounts.) On the far right, if you have any IM conversations in progress, you see an indicator (such as "2 chats"). (When you get up to 8 or 10 chats, you can wave your G1 around and win any popularity contest that comes along.)

Setting Up Accounts

In the case of Google Talk, your account is already set up for you; it's part of your Google account, which your G1 has known since you went through the initial setup process. Setting up AIM, Windows Live Messenger, and Yahoo! Messenger accounts is a snap, though.

If you don't have an AIM account but do have an AOL e-mail address, I have good news: You actually *do* have an AIM account after all. Just use your AOL e-mail address for your username and your e-mail account's password when setting up your AIM account in the following steps. Similarly, if you have a Hotmail e-mail address, it can be used as your Windows Live Messenger account, and Yahoo! Mail accounts can be used as Yahoo! Messenger accounts.

Let's say you want to set up your AIM account on your G1. Starting from the account selection screen:

1. Touch the AIM row.

The Add AIM account screen appears, as shown in Figure 12-2.

Figure 12-2:
Adding a
new AIM
account.

2. **Using the keyboard, fill out your screen name and password for the AIM service.**

   If you don't already have an AIM account, you can set one up right from your phone — just touch the Get a Screen Name link at the bottom of the screen.

3. **If you want the phone to remember your password so that you don't need to type it each time you sign in, touch the Remember my password check box to select it.**

   A reminder appears on-screen with some sound advice: Should your G1 ever be lost or stolen (and I certainly hope it isn't), immediately use the AIM Web site to change your password so the bandit can't sign in with your identity.

4. **If you opt to let the phone remember your password, you can select the Sign me in automatically check box to keep yourself signed in on your phone at all times.**

5. **Touch the Sign in button.**

And that's it! After a moment, you're brought to your contact list for this account, which I get into later in the chapter, in the "Contact Lists" section.

You add your Windows Live Messenger and Yahoo! Messenger accounts in the same way as you add AIM. The only difference is that Windows Live Messenger requires an e-mail address in place of a screen name and also asks you to agree to terms of use before allowing you to proceed — touch I accept to continue. (You have to deal with this prompt only once.) Yahoo! Messenger, meanwhile, will prompt you for a username in place of a screen name, but the concept is the same.

Managing Accounts

Once you have your IM accounts set up, you use the account selection screen (refer to Figure 12-1) as your jumping-off point to manage and use each account. If you're already signed in to a particular account, simply touching its row in this screen's list will take you to your contact list for that account. If you're not signed in, touching the row will first sign you in before taking you to the list. (You'll first be prompted for your password if you decided not to let the G1 save it when you set up the account.)

Touching and holding on a row brings up one of three pop-up menus. If you have an account set up and you are signed in, you see a menu like the one in Figure 12-3.

Figure 12-3:
The menu
for a
signed-in IM
account.

Touching Contact List takes you to the contact list for that account. This option performs the same function as simply touching the account's row. Sign out takes the account offline; you cannot send or receive instant messages on the account, and other users of the same service will not see you in their contact lists. Touching Settings brings you to the Settings screen for the account; I discuss your options there in depth later in the chapter, in the "Instant Message Settings Screens" section.

If you call up the pop-up menu for an account that's not currently signed in, you see the menu in Figure 12-4 instead.

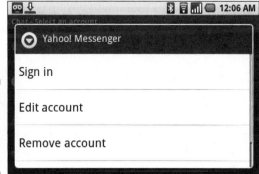

Figure 12-4:
The menu
for a signed-
out IM
account.

Here, you're presented with four options:

✔ **Sign in:** The first option signs you into your account (prompting you first for your password if it's not saved) and brings up your contact list. This option is the same as touching the account's row in the list if you were already signed in.

✔ **Edit account:** The next option brings you back to the screen you see when creating your account for the first time, allowing you to edit the username (or screen name or e-mail address, depending on the service) and password and set the password saving and auto sign-in options.

✔ **Remove account:** Use this option to remove the account you've configured for this service. If you touch this service after the account has been removed, you'll be prompted to enter new account information from scratch.

✔ **Settings:** This option takes you to the Settings screen for the service.

Finally, if you touch and hold on a service for which you don't have an account configured, the pop-up menu offers you just one option: Add account. This option performs the same function as touching a service name: You go to the account creation screen, where you configure your screen name (or username or e-mail address) and password.

Although you can touch and hold on each signed-in account individually to sign out, many times you'll probably want to sign out of *all* your accounts at once. No problem — just push Menu while in the account selection screen and touch Sign out all.

Contact Lists

Each of the four services has its own *contact list* — a list of the people with whom you can interact via instant message. If you've added accounts to your G1 that you already use on your PC, those contact lists have conveniently been transferred; otherwise, you start with a clean slate.

AIM, Windows Live Messenger, and Yahoo! Messenger

The contact lists for all three of the non-Google IM services supported by the IM application work in essentially the same way, so I'll cover them under the same umbrella. Figure 12-5 shows an example from AIM.

At the top of the screen, a gray bar shows you the currently logged-in AIM screen name (in this case, "tmobileg1fd"). Directly below that, your current status is shown — this is the status other users see so they know whether you're available, busy, or away (the exact verbiage of the status messages varies from service to service, but the concept is the same).

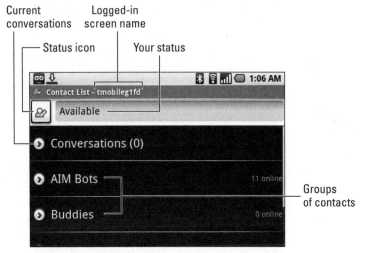

Current
conversations

Logged-in
screen name

Status icon

Your status

Figure 12-5:
The AIM
contact list.

To change your current status:

1. Touch the status icon in the upper left of the screen.

In this example, I have my status set to Available, which is spelled out to
the right of the status icon.

2. Select the status you want to set.

For AIM, the available choices are Available, Away, and Appear offline;
each status has its own icon, which you see in the upper left of the
contact list after you select one.

Google Talk and Yahoo! Messenger also allow you to set a textual representa-
tion of your status. For example, Yahoo! Messenger's default text for the status
indicating you're available is Available (what else?), but you can modify this to
whatever you'd like — other users will then see this text in their contact lists.
To change this text, touch the text box to the right of the status icon, edit the
text to your liking using the keyboard, and press the Enter key or touch any
other part of the screen to confirm it.

Below the status area, your contacts appear, organized into one or more
groups. (You're not stuck with the default groups provided when you set up
an account, but you do need to use each service's PC application to modify
them.) These groups can be expanded or collapsed by touching their rows.

The topmost group isn't actually a group of contacts at all; it's a special holding area for conversations that you currently have underway. I discuss conversations more (after all, conversations are what instant messaging is all about!), in the "Managing Conversations" section.

By expanding a group and touching and holding on a contact, you are presented with a pop-up menu of commands you can perform on that contact. The contents of the menu vary depending on the service — for example, AIM calls chats IMs but Yahoo! Messenger simply calls them chats — but the concepts are the same:

✔ **Send IM/Start chat:** This option initiates an instant message conversation with this contact.

✔ **Buddy info/View profile:** Touching this item brings up a window with more information about this contact (see Figure 12-6). Depending on what information the service makes available to you, you may see a contact's image, e-mail address, username, screen name, and current status (Available or Away, for example) and a client type, which simply lets you know whether the contact is currently using a computer or a mobile device such as a G1. To exit from this window and return to the contact list, push the Back button.

✔ **Block:** If the service you're using supports the blocking of contacts, this is where you do it. Blocked contacts are unable to contact you — great for that pestering coworker (used in moderation, of course).

✔ **Delete Buddy/Delete contact:** Use this option to delete the contact from your contact list. Touch OK in the confirmation window to go ahead with the deletion or Cancel to back out.

Figure 12-6:
AIM's
Buddy Info
window.

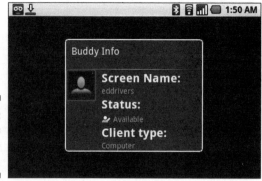

Adding a contact

To add a contact to your list, first select the group to which you want to add your new contact. Then:

1. **Push the Menu button.**

2. **Select the Add Buddy (or Add contact, depending on the service) menu item.**

   The Add Buddy/Add contact screen appears.

3. **Type the screen name, username, or e-mail address of the contact you're looking to add.**

   As you type, the application searches your contacts looking for matches and offers any suggestions that it can find. If you see one that matches what you're looking for, simply select it.

4. **Touch Add Buddy (or Send invitation).**

Congratulations — you've just added a contact to your list! Some services and accounts require approval by the recipient before you're allowed to add them to your contact list; if that's the case, you may not see the contact right away.

Viewing blocked contacts (and unblocking them, if they're lucky)

By pushing Menu and selecting the Blocked menu item, you'll be taken to a screen that shows you all contacts that you currently have blocked on this account (see Figure 12-7).

From this screen, simply touch a contact to unblock him or her. You'll be asked for confirmation; touch OK to proceed with the unblocking or Cancel if you want to let 'em stew a bit longer.

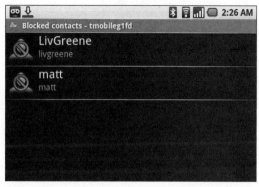

Figure 12-7: Your list of blocked contacts.

Google Talk

Gmail has its own e-mail application on the G1 that features some bonus functionality not found in Email, its non-Google counterpart, and Google Talk gets similar treatment! In this case, Google Talk is still part of the same IM application used by the other major services, but it works and acts a little differently.

The Friends list (see Figure 12-8) is the home base for all your action-packed Google Talk operations. The screen should look familiar if you've set up AIM, Windows Live Messenger, or Yahoo! Messenger, with a few key changes.

First off, you'll notice an additional picture to the left of the status icon at the top. This is your personal Google contact icon. By touching it, you can replace it with a picture on your G1, which you'll have an opportunity to crop into a square before saving.

Just like the other IM services, touching the status icon allows you to set your status, and you can also type a textual status message into the text box to the right. Below that, your contacts appear. In Google Talk, contacts aren't organized into groups — all your contacts appear in a list, with active conversations at the top.

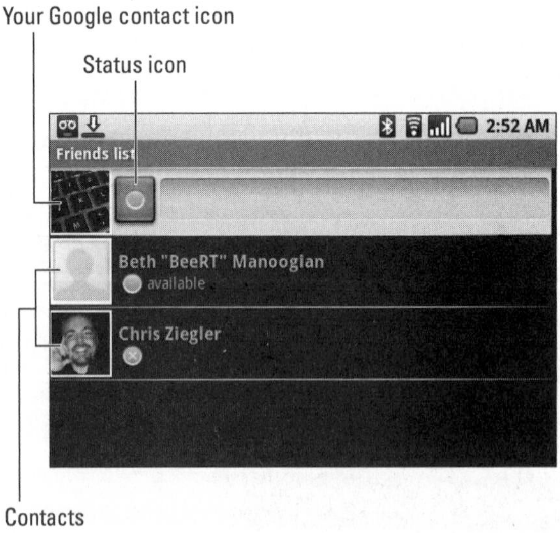

Your Google contact icon

Status icon

Figure 12-8: The Google Talk Friends list.

Contacts

If a contact in your list is currently connected to Google Talk using an Android-powered device such as a G1, you'll also see a small Android logo to the right of the contact's name.

You can get to your most-contacted contacts quickly by pushing Menu and choosing the Most popular menu item, which filters out folks you don't chat with much. To return to the list of all contacts in your Google Talk list, push Menu again and choose All friends.

Adding a contact

Adding a contact with Google Talk is a lot like adding a contact in one of the other IM services but even easier, because all you need to know is an e-mail address — there are no separate usernames or screen names to worry about. Push the Menu button, select Add friend, and type the e-mail address of the person you want to add. (The application will make suggestions from your contacts as you type.) When you're finished, touch Send invitation, and the contact you want to add receives a notice that you're trying to add him or her to your contact list. If the person approves the request, he or she shows up in your contact list.

Feeling loved: When a contact wants to add you

The reverse holds true, too: When people want to add you to their contact list, you'll receive an invitation. When this happens, you get a notification in your status bar and a message in your Google Talk contact list (see Figure 12-9).

To accept or decline (ouch!) the invitation, just touch it in your contacts list and select the appropriate option, as shown in Figure 12-10.

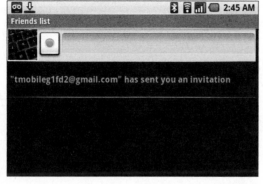

Figure 12-9:
A new
Google Talk
invitation.

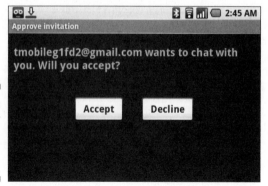

Managing Conversations

Once you have your contacts situated, you're ready to instant message (and be instant messaged) like a pro. To initiate an IM conversation with someone, just touch the person's name in the contact list of any of your services — keep in mind you have to be signed in first — and you are brought into the conversation view (see Figure 12-11).

Sending new messages works much like sending a new SMS message from the Messaging application — just type your message in the box at the bottom of the screen and press Enter on the keyboard or touch Send on the screen.

As you receive messages, you'll be notified in a couple of ways. Most importantly, you receive a notification in the status bar (unless you've turned off notifications in Settings, which I cover later in the chapter, in the "Instant Messaging Settings Screens" section). When a notification arrives, you can pull down the notifications screen (see Figure 12-12) and touch the new notification to go directly to the conversation.

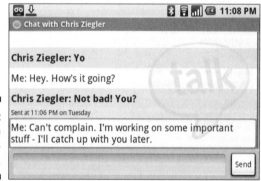

Figure 12-11:
An active
conversa-
tion.

For services other than Google Talk, the conversation appears also in your group of active conversations, which is at the top of each service's contact list. In the case of Google Talk, you simply see the contact change to a "bubble" within your contact list, as shown in Figure 12-13, to indicate that you now have an active conversation with the contact.

Pushing the Menu button while in an active conversation brings up a slightly different menu depending on the service you're using. Some of these commands are duplicates of commands found elsewhere in the application (such as Block), but others bear explanation here. Let's have a look at the highlights:

- ✔ **Contact list/Friends list:** This option returns you to your contact list. You can also push the Back button to accomplish the same function. (Choices are good, right?)

- ✔ **Switch chats:** When you have multiple conversations going at once, it's convenient to be able to switch between them without first having to go back to your contact list and touch another contact. The Switch chats menu item lets you do just that. Touching it brings up a box (see Figure 12-14) with a list of all your contacts (regardless of service) with whom you presently have an active conversation; just touch a contact to flip to that contact's conversation screen.

- ✔ **Add to chat:** With Google Talk, you can get a group chat going — a conversation between multiple contacts at once. (Think of it as a textual party line.) Touching this option brings up a list of contacts you can add, and touching a contact in the list sends the person an invitation to join the chat.

- ✔ **Insert smiley:** These days, we've gone far beyond the basic emoticons like :-) and :-(. Now we have to worry about :-X, B-), and o_O too! Fortunately, you can use this option to select from the IM application's library of emoticons, so there's no need to commit them all to memory. The list of available emoticons varies because each service's PC application can automatically translate certain emoticons to images — turning :-) into an actual image of a smiley face, for example — so only the ones that can be translated by that service are displayed.

Figure 12-14:
The Switch chats window.

Instant Messaging Settings Screens

Pushing Menu from within any service's contact list screen will give you access to that particular service's Settings. You can access them also by touching and holding the service's row in the account selection screen and choosing Settings. AIM, Windows Live Messenger, and Yahoo! Messenger all share identical Settings screens, while Google Talk's is slightly different. I explain those differences as I walk through all the items here:

- ✔ **Automatically sign in:** This option is available only in Google Talk. Unlike the other services, Google Talk has no account setup screen (your account is built into the G1 as soon as you set it up for the first time). You use this option to configure automatic sign-in for the service. It simply keeps you signed in, if you dare make yourself available via instant message to ravenous friends, family, and coworkers at all times!

- ✔ **Mobile indicator:** This menu option is available only in Google Talk. When you select this option, your contacts can see that you're using a phone to communicate with them, which lets them know that they should expect slightly slower responses than if you were on your PC. (The phone could be in your pocket, after all, and most of us aren't as fast typing with our thumbs as we are with our full hands.)

- ✔ **Hide offline contacts:** This option is not available in Google Talk. This item hides your contacts that aren't currently online from your contact list, so you don't end up with a jumbled, overcrowded list of people you can't contact. Good for heavy IM users with lots of contacts.

- ✔ **IM notifications:** This option enables notifications of new instant messages via Android's notification system. In other words, you'll see the messages in the status bar and the pull-down notifications screen. This option is enabled by default, and it's highly recommended; otherwise, knowing when you've received a new IM can be difficult.

- ✔ **Vibrate:** When this option is enabled, the phone briefly vibrates when you receive a new instant message.

- ✔ **Sound:** This option is not available in Google Talk; you always hear a sound for Google Talk messages unless the phone is in Silent mode. When this option is selected, the service plays a sound when you receive an instant message.

- ✔ **Select ringtone:** This option allows you to select the sound that is played when an instant message is received (bear in mind that this option has no effect if you deselect the Sound check box). Touching this item brings up a menu of available sounds; select a sound to hear a preview. When you find a sound you like, touch OK to confirm your selection. Touch Cancel to revert to the previous selection.

When you add in four major services of instant messaging to the G1's huge spread of communications tools, you'll find that you're as connected (heck, maybe more connected) from the palm of your hand than you are in front of your PC. Just make sure to do yourself a favor and try to keep 'em all turned off when you're trying to catch a little R&R on the beach, okay? You'll thank me later!

Chapter 13

Rock Out with the G1's Music Player

. . . .

In This Chapter

▶ Choosing headphones versus a loudspeaker

▶ Browsing and searching your music collection

▶ Managing your playlists

▶ Using songs as ringtones

▶ Working in the Playback screen

. . . .

*T*he G1 includes a memory card slot that supports the latest generation of high-capacity microSD (commonly known as microSDHC — try saying *that* five times fast) cards, so the phone can swallow 16GB of storage capacity or more. Elephants may never forget, but even the mightiest pachyderms don't hold a candle to a well-equipped G1 for storing all the music that's important to you. (The G1's a whole lot more portable, too.)

Your G1 may have come with a 1GB microSD card already installed. For many music aficionados, this isn't enough. Cards as large as 12GB or 16GB are readily available from local electronics stores and on the Internet for under $100 — a wise and easy investment that helps turn your phone into the all-in-one media powerhouse that it's capable of becoming. I discuss the process of removing and installing memory cards in Chapter 15. (Don't worry — it is a snap and takes only about 15 seconds).

Once you have a nice, fat memory card installed and you're ready to rock, you can get music into your G1 in several ways. In Chapter 16, I describe Amazon MP3, the G1's built-in music store where you can purchase music and then install it anywhere you can pick up a Wi-Fi signal. If you already have a ton of music on your PC that you're ready to add, you may want to consult Chapter 17, where I talk about connecting the G1 to your computer to download (and upload) music and more.

That said, the G1 isn't the perfect portable music player. A glaring deficiency is its lack of a 3.5mm headphone jack, also known as a $1/8$-inch jack or a mini jack — the universal standard for connecting headphones in modern players. Instead, the phone offers audio through its ExtUSB jack, so you have to use the special headphones that were included (which aren't going to fit the bill if you're picky about sound quality) or use a separate 3.5mm adapter. The adapter's just one more thing to carry around — and one more thing to lose.

Android's built-in music player probably isn't the most full-featured you've ever used; it's a back-to-basics application that just gets the job done with a few frills thrown in for good measure. Of course, one of the great things about Android is that nothing is stopping you from downloading a new, different, better player from the Market (which I cover in Chapter 18) or another source, but that doesn't overcome the G1's lack of a 3.5mm jack. Ultimately, I like to look at the G1 as an "in case of emergency" music player; it gets the job done, but if you're the type of person who is hard to catch without a pair of earbuds on your face, you might be better off carrying a separate dedicated device (such as an iPod or a Zune) as well.

In this chapter, you find out how to maximize the G1's Music application, discovering a few tricks along the way. By the end of this little journey, you'll be rocking out with the best of 'em.

Choose Your Destiny: Headphones or Loudspeaker

Normally you'll want to keep your funky beats to yourself, which is why you have headphones. (Well, okay, maybe you'd rather share your tunes with the world, but the rest of us enjoy our peace and quiet, thank you very much!) Occasionally, though, you might find that the headphones aren't convenient, you don't have them with you, or you just have to share that new Celine Dion track with a friend. Fortunately, the G1 gives you a pretty decent way to do that.

If you plug in headphones, music is automatically piped through them and you won't hear anything coming from the G1 itself. If you don't plug in the headphones, however, the G1 automatically knows to play music over its loudspeaker — the same speaker it uses for the ringer and speakerphone.

If you turn the phone over so its back side is facing toward you, you'll notice a tiny bump right below the speaker grate. This bump was ingeniously included so that the speaker wouldn't be muffled by placing the phone face-up on a surface. It's most handy for the ringer so that you can actually hear when someone is calling you after you've set your phone down, but it serves the convenient dual purpose of making music audible, too.

You'll likely find that the loudspeaker mode isn't helpful unless you're in a very quiet room, and even then, you'll want the phone at or near the maximum volume setting. That said, though, the sound is surprisingly clear. It's no boom box, but it'll take care of business in a pinch.

Starting the Music Application

If you're not already playing music, you can start the Music application the same way you would any other app: just navigate to the Applications tab from the Home screen and touch its icon (or touch the icon on the Home screen itself, if you've added a shortcut there).

If music is currently playing, however, there's a second way to start the Music app, which I get to in a moment. But first, a quick explanation of the Music application's use of Android's notification system is in order. The application cleverly makes use of the notification system to help you keep tabs on what's playing right now — a handy shortcut for checking a track name without having to go all the way into the app. When a track is playing, you see a play symbol in the status bar, as shown in Figure 13-1.

Play symbol is visible when a track is playing

Figure 13-1:
The Music
app's status
bar icon.

Pulling down the notifications screen (see Figure 13-2) reveals the name of the track, the artist, and the album in a special section of the screen labeled Ongoing. These aren't one-time notifications such as a new text message or e-mail. Instead, they remain in the notifications screen until a certain condition has changed. In this case, the current condition is that music is playing, so you'll see this notification until the music stops.

I promised that I'd reveal a second way to open the Music application — lo and behold, here it is. Simply touch the notification here to open it, and you are taken straight to the Playback screen to see details about the track and control playback.

Figure 13-2:
The notifica-
tions screen
shows
current
track
information.

Browsing and Searching Your Music

Opening the Music application for the first time will take you to the Library screen (see Figure 13-3), which you can think of as the home base for your music browsing and searching activities. From here, you can view lists of music on your G1 organized by artist or album, see a listing of all tracks, or check out your playlists.

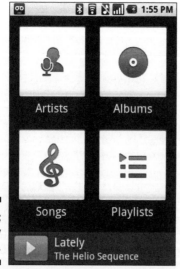

Figure 13-3:
The Library
screen.

As when using iTunes, an iPod, or any other software or hardware designed to play digital music tracks, accurate and complete track information is important! The G1 fully and correctly organizes each track only when its artist name, album name, year, album artwork, and track number are filled out. The Music application will do its best to work with incomplete information, but if you find a lot of "Unknown artist" or "Unknown album" labels while playing your tunes, you're best off spending some time filling out that information using the music player on your PC. Many modern software players (including iTunes) can analyze your music and attempt to fill in missing information, so using a PC-based music application first is a good time-saver to try before editing information yourself.

Pushing Menu when in the Artists, Albums, or Songs screens will give you similar choices:

- ✔ **Library:** The top option returns you to the Library screen, where you can select the Artists, Albums, Songs, or Playlists screens.

- ✔ **Playback:** This command opens the Playback screen — some software and media players refer to this as the Now Playing screen. I talk more about everything this screen can do later in the chapter, in the section titled "The Playback Screen."

- ✔ **Shuffle all:** Choose this option to shuffle everything you currently see into one big randomly generated playlist.

- ✔ **Play all:** The Songs screen offers this additional menu item, which simply plays all songs in alphabetical order; no shuffling is applied.

Browsing artists

To browse your music collection by artist, touch the white box labeled Artists or select it with the trackball (an orange border surrounds the box when it's selected). This brings you to the Artists screen, shown in Figure 13-4.

Each row represents an artist whose music is loaded into the phone. Below the artist's name, you see the number of complete albums loaded (if any) and the total number of songs by that artist loaded. The rows start out in the collapsed state (indicated by a right-facing arrow to the left of the artist's name). By touching a row, you expand it, revealing the albums available to you.

Even if you have loaded only a partial album (or just a single song from an album) the G1 still organizes the tracks into the albums on which they appear.

Collapsed row

Expanded row

Album name

Number of tracks
you have loaded

Year album was released

Figure 13-4:
The Artists
screen with
an artist
expanded.

Each album (or partial album) listed under an expanded artist displays the album name, the number of tracks you have loaded on your G1 from that album, and the year the album was released (if the G1 can gather that information from the track data). You also see a thumbnail view of the album artwork to the left if it's available; if not, you'll see a generic image of a CD.

If a song is currently loaded in the Playback screen — regardless of whether it's paused or playing — you see an orange play icon (a forward arrow) in the lower-right corner of that artist's or album's row.

Touching an individual album's row takes you to the track listing for that album. This screen operates in the same way as when you're browsing all songs on your G1 (see the "Browsing songs" section later in the chapter), except you see only tracks on the specific album you selected. Additionally, the tracks are presented in correct album order (instead of alphabetically) if the track numbers are present in the files.

Touching and holding on an artist or an album in Artists screen displays the same pop-up menu. Let's take a look at the items you'll find there:

✔ **Play:** If this menu item is selected for an artist, all albums by that artist are played in alphabetical order, and the tracks within each album are played in the order in which they appear on the album. Touching this option while an individual album is selected plays just that album's tracks.

✔ **Add to playlist:** This option adds the artist's or album's tracks to the current playlist, to a playlist you've previously saved, or to a new one. I touch on playlists later in the chapter.

✔ **Delete:** The final option removes all the tracks for the selected artist or album from the phone's memory. In the confirmation window that appears, select OK to continue with the deletion or Cancel to back out.

Browsing albums

To browse all albums on your phone, go to the Library screen (you can return to it from anywhere in the Music application by pushing Menu and selecting the Library menu item) and select the white box labeled Albums. You see a screen that looks like Figure 13-5.

The function of this screen is simple: It gives you a list of all available albums (and partial albums). Each album is represented by a row that tells you the album's name at the top, followed by the artist and the number of songs from that album that you have on the G1 — this number will not be the total number of tracks on the album if you haven't loaded the entire album onto the phone.

Figure 13-5:
The Albums screen.

Touching an album takes you to its full track listing — the same screen you see when you select an album from the Artists screen. Similarly, you get the same pop-up menu you're already familiar with from the Artists screen by touching and holding on any of the albums; from this menu, you can play the entire album in its correct track order, add it to your current or a new playlist, and delete the album from your G1.

Browsing songs

If you're looking for a simple list of every track you have on your phone sorted in alphabetical order, the Songs screen shown in Figure 13-6 is your new best friend. To get to it, touch or select the white box labeled Songs from the Library screen.

For each track, you see its name, album name, and artist name on the left, accompanied on the right by the duration of the track in minutes and seconds. From here, you can simply touch or select a track to play it. Touching and holding on a track displays a pop-up menu with the following items:

Artist's name

Album name

Track name

Track duration

Figure 13-6:
The Songs
screen.

- ✔ **Play:** This option simply plays the selected track, which is the same as touching the track's row.

- ✔ **Add to playlist:** This option adds the track to the current playlist, a saved playlist, or a new playlist that can be saved.

- ✔ **Use as phone ringtone:** Select this option to use the selected track as your default ringtone for received calls. A cool, quick, easy way to hear your favorite song every time someone dials you up!

- ✔ **Delete:** When you want to remove the track from your phone's memory, select this option. You'll be presented with a confirmation window. Touch OK to go ahead with the deletion or Cancel to back out.

Searching your music

If one of those expansive 16GB microSDHC cards is installed in your G1, you have a ton of space to add music — but the downside is that it becomes harder and harder to browse through lists of artists, albums, or individual tracks to find something. To put things in perspective, if you were to fill a 16GB card to capacity, you could have well over 3,000 songs!

Well, no Google-created application would be complete without incredible searching capability, would it? Of course not, and the Music application is no exception.

Say you're looking for the group Flight of the Conchords. Here's what you'd do:

1. **From wherever you happen to be in the Music app, push Menu.**

2. **Select the Library menu item.**

   This returns you to the Library screen.

3. **Push Menu again and select Search.**

   Alternatively, if you already have the keyboard open, you can simply press the Search key without pushing Menu. (You need the keyboard open anyway to complete the next step.)

   You are taken to the Search screen. This is an interesting view of your music — artists, albums, and individual songs are lumped together into a single, huge, searchable list.

4. **Start typing** flig.

As you type (see Figure 13-7), each letter appears to float in the lower center area of the screen, and the Music application searches through all your artists, album, and individual track names, looking for words that match what you've typed so far. Anything that doesn't match is filtered out, so as you type more letters, the list becomes more and more filtered until finally only the artist, album, or track you're looking for is displayed. If you make a mistake, just use the Backspace key as you would in any text field in Android.

Note that an icon appears on the left side of each row — this indicates whether that row represents an artist (a silhouette of a person behind a microphone), an album (a CD icon), or a track (a treble clef).

5. **When you see what you're looking for, touch it or select it with the trackball.**

Depending on what you touch, a different action occurs:

- **An artist:** You see a list view of all albums and partial albums by that artist.

- **An album:** You see a list view of the tracks on that album.

- **A track:** The track immediately begins playing.

Figure 13-7:
Searching
for Flight
of the
Conchords
from the
Search
screen.

Managing Your Playlists

If you plan on listening to more than a single track or two, odds are you'd like your music player to do the heavy lifting of deciding what track is going to play next — otherwise, you'll be going into the app every five minutes to select the next song. Where's the fun in that? The G1's Music application includes plenty of options for creating and maintaining playlists that should keep the beat going for as long as you need it to.

The Playlists screen

You can see all playlists saved on your G1 by going to the Playlists screen, which is accessible from the Library screen by selecting the white box labeled Playlists.

To start playing a playlist, just touch its row. The first track in the playlist starts playing, and as each track ends, the Music application selects the next track in the playlist and starts playing it.

When you first get started using the Music application, you won't have created any playlists (but don't feel bad — we'll get to that in just a moment). The only thing you'll see here is an automatic playlist called Recently added that the Music application generates and maintains for you.

The Recently added playlist is simply a playlist of all the songs that have been recently added to your G1. Just bought and loaded a few albums that you're excited to hear? Recently added is the best place to do it.

By default, Recently added includes music that you've added in the past two weeks, but you can adjust this. To do so:

1. **Touch and hold on the Recently added item until a pop-up menu appears.**

   You see two options, Play and Edit. Play starts playing the playlist, just as touching the playlist's row does.

2. **Select the Edit menu item.**

   You see a list like the one shown in Figure 13-8, where you can select how many weeks back from today you want included in the Recently added playlist. (You can choose anywhere from 1 to 5 weeks.)

Figure 13-8:
The adjustable duration of the Recently added playlist.

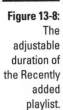
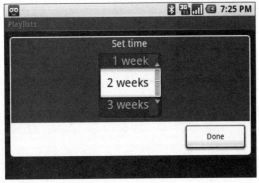

Creating a playlist

You can create a new playlist in several ways, each of which you may find useful in different situations:

- **From a track, an album, or an artist in a list:** Whenever you're looking at a list of tracks, albums, or artists, you can touch and hold on an item to get a pop-up menu and select Add to playlist. From the window that appears, select New to indicate that you want to create a new playlist. You are prompted for the name of the playlist, as in Figure 13-9 ("Totally awesome ABBA tracks," for instance). Press OK to add your saved playlists for future use.

- **From an automatically generated playlist:** Say you've decided to shuffle all the tracks on your G1 using the Shuffle all menu item in the Songs screen, and you *really* like the way they've been shuffled. Wouldn't it be nice if you could save this playlist permanently for posterity? You can! Pushing Menu while looking at the current playlist reveals a Save as playlist menu item, which when pressed prompts you for a name for the new playlist. Touching OK commits the playlist to your list in the Playlists screen.

- **From the Playback screen:** Pushing Menu from the Playback screen gives you an Add to playlist option. Selecting this item and selecting New in the next window that appears adds the currently playing track to a new playlist with a name of your choosing.

Figure 13-9: Choosing the name of a new playlist.

Adding music to an existing playlist

In the process of trying out the different ways to create a playlist, you may have noticed something interesting: In addition to creating a playlist by selecting New, you had the option of adding the selected item to the current playlist or any of the other playlists you've previously saved.

If you touch and hold on a track, an album, or an artist and select Current playlist from the Add to playlist pop-up menu, the playlist that's currently playing will have the selected track, album, or artist tacked on to the end of it. This is handy if you just need a few hours of music right now and none of your playlists matches your mood — you can start some music playing while continuing to work on the playlist. It's *almost* like you're a DJ, except you're working with one G1 instead of two turntables.

If you select the name of one of the playlists you've previously saved, the selected item will go to the selected playlist instead. If you're in the process of creating a playlist out of individual tracks from different artists and albums, you'll want to select New to create a new named playlist the first time you select a track; on each subsequent selection, you select the name of the playlist that you just created.

Editing a playlist

From the Playlists screen, you can delete existing playlists, rename them, and rearrange and delete individual tracks within them. Return to the Playlists screen and touch and hold on a playlist that you've created (not Recently added). You get a pop-up menu with four menu items:

- ✔ **Play:** This option simply plays the selected playlist — the same as touching it.
- ✔ **Delete:** This command deletes the playlist from your phone.

   Deleting a playlist doesn't delete the music that's in the playlist — it deletes just the playlist itself.
- ✔ **Edit:** Touch this option to rearrange tracks in or delete tracks from a playlist.
- ✔ **Rename:** Use this option to rename a playlist.

Let's take a closer look at the Edit option. Touch and hold on a playlist that you've created and choose Edit. You see a screen similar to Figure 13-10.

Notice the "grabbers" that appear to the left of each track. By touching these circular icons and holding down, you can move tracks around to reorder them — it's that easy. To remove a track altogether, touch and hold its row and choose Remove from playlist from the pop-up menu that appears.

Ringtones, the law, and you

With the amount of press swirling around the RIAA (Recording Industry Association of America) and its pursuit of folks illegally downloading and distributing music, it's fair to ask whether turning a track you legally purchased (or otherwise acquired) into a ringtone is within the bounds of the law. The short answer is "yes." A recent court ruling determined that ringtones aren't considered derivative works of an artist, which would be subject to additional royalties. Some contention exists as to whether ringtones constitute a public performance of a song — another contingency that would demand that more money be paid to the publisher — but the Copyright Office's current stance is that they're for private use, even though others can hear them.

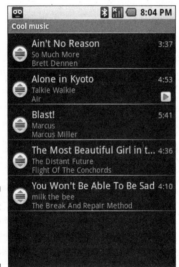

Figure 13-10: Editing the tracks in a playlist.

The Playback Screen

As you may have noticed from earlier in the chapter, there's no shortage of ways to get into your Playback screen. You can get there from the menus of the Artists, Albums, Songs, and Playback screens; from the notifications screen when a track is currently playing; and also from a gray bar at the bottom of the Library screen that lists the currently loaded track and artist name. It may seem like overkill to provide so many ways to get to a single screen in a single application, but the Playback screen is pretty freakin' important: It's how you control what you're listening to, after all!

When you arrive at the Playback screen (see Figure 13-11), you're presented with a wealth of information and options, so let's break down everything you're seeing here.

At the upper left, the album artwork for this track appears (it's simply a larger version of the thumbnail you see in the Albums screen). To its right, three icons appear, in the following order:

- ✓ **Now playing:** The first icon shows you the playlist that's currently playing. Touching this icon takes you to a playlist editing screen where you can see what tracks are coming up next, rearrange and remove tracks, and save the playlist to your Playlists screen (this is handy, for example, if you're listening to a shuffled mix that isn't already saved as a permanent playlist). When you're in this view, the track that's playing right now is indicated with an orange play icon in the lower-right corner of the row.

- ✓ **Shuffle:** When you touch this icon, it turns from gray to green, and you receive a message on the screen indicating that Shuffle is on. This option shuffles the tracks in the playlist you're currently playing. To disable Shuffle mode, touch the icon again.

- ✓ **Repeat:** Touching this icon the first time repeats all songs in the playlist; that is, when the Music application reaches the end of the playlist, it loops back to the beginning and starts over. Touching the icon a second time repeats only the song that's currently playing (so it better be a good one). A third touch turns off Repeat mode.

One track backward

Now playing

Shuffle

Repeat

One track forward

Total track length

Current track position — Play/Pause

Figure 13-11: The Playback screen.

Below the album artwork, you see the artist, album, and track name, presented in that order.

Next, icons to the left and right allow you to move one track backward and forward in your current playlist, while the icon in the center plays and pauses (the icon changes to pause when you're playing, and vice versa). The time to the left is the current position in the track, and the time to the right is the total length of the track.

Below those controls, a bar visually indicates how far you are through the track. By touching any position on the bar, the track will jump to that position. You can also drag your finger back and forth along the bar to "scrub" the track — if the music is playing, you continue to hear the current position of the track as you move the slider. It's a handy feature for quickly (and audibly) finding the location of a song or audiobook that you want to play.

Want to find out more about that great jam you're currently listening to? The Music app makes it easy. Just touch and hold on the artist, album, or track name, and after a moment you'll see a pop-up menu (see Figure 13-12) that lets you search for the name in the Amazon MP3 music store, on the Web using Google search, or in YouTube. These options are great ways to get more music, info, and videos on that hot new musician.

Figure 13-12:
The G1
makes
searching
for more
music, info,
and videos
by your
favorite (and
most hated)
artists a
breeze.

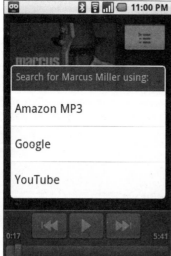

Pushing the Menu button from the Playback screen will give you a handful of options that'll come in handy at one time or another:

- **Party shuffle:** This is really just a fancy way of saying "take all the tracks on my G1 and mix 'em up." By selecting this menu item, the Shuffle icon toward the upper right of the Playback screen changes from two interchanging arrows to a mirror ball (yes, just like the kind they use in disco halls), and the first track immediately begins playing. You can turn off Party shuffle by touching the Shuffle icon or the Party shuffle menu item again.

- **Add to playlist:** As I mentioned earlier in the chapter, one of the places where you can create a new playlist is from right here in the Playback screen. Touching this menu item will let you add the playing track to the current playlist (although I'm not sure why you'd want to do that, since it's already *in* the current playlist), to a new playlist, or to one of your saved playlists.

- **Use as ringtone:** Like what you hear? Why not hear it every time some-one calls? Just touch this menu item and the current song becomes your default ringtone.

- **Delete:** This option deletes the song that you're currently listening to from memory.

If you've been following along in this chapter and testing out the Music app on your G1 as you go, there's a good chance your ears are ringing by now — good thing you only have to *read* this book, isn't it? When you're ready to put Metallica on pause for a few minutes, follow me into Chapter 14, where we look at some of Android's smaller (but no less important!) applications.

Chapter 14

Best of the Rest: Alarm Clock, Calculator, and YouTube

*O*dds are pretty good that you'll spend most of your precious G1 time using heavy-hitting applications such as Contacts, Calendar, Maps, and the Browser, but Google has thrown in a supporting cast of smaller programs that deserve an honorable mention.

You may use the Alarm Clock and Calculator applications only a few times a year but when you need them, you'll be really glad they're there. All it takes is one night in a hotel room with a clock radio that's impossible to decipher (or even worse, busted) to make the G1's integrated Alarm Clock app pay dividends.

Without any built-in games, the YouTube application is the G1's first, last, and only line of defense against boredom (until you start downloading your own applications — see Chapter 18 for that). Plus, YouTube is a Google-owned service, so it only makes sense that it'd want to feature one of the world's most popular Internet video services front and center on Android.

In this chapter, I tell you everything you need to know about setting up the Alarm Clock, running the numbers with the Calculator, and killing a few minutes with YouTube. (And yes, the material is presented in that order — you have to eat your dinner before you can have dessert!)

Alarm Clock

Think an alarm clock is just a time that you set for your phone to beep at you? Think again. Well, actually it is — but in the case of Android's Alarm Clock application (see Figure 14-1), it's also quite a bit more. This program gets a little fancier than your average bedside aid by letting you set as many alarms as you want, choosing different sounds for each and setting them to be active only on certain days of the week.

To start the Alarm Clock, choose its icon from the Applications tab. At the top of the screen, you see a large clock — not quite as pretty as the clock widget on Android's Home screen, but it gets the job done. Below that, you see a list of all programmed alarms (if you have none, you'll see an empty black space here). On the left, the time of the alarm is indicated; if it's set to be active only on certain days of the week, you see a listing of those days below the time.

 Just because an alarm is programmed doesn't mean it's active! To the right side of each alarm's row, a green check mark in the check box indicates that the alarm is set to go off. Touch the check box to select or deselect it.

 When you move around the world with your G1 in tow, the phone will automatically change its clock and time zone (as most cellphones do). Although entries in your Calendar will automatically change time to reflect the time zone change, alarms will not. In other words, if you have an alarm set for 8 a.m., it will still be set for 8 a.m. in your local time no matter where you go.

Programmed alarms   Indicates an alarm is active

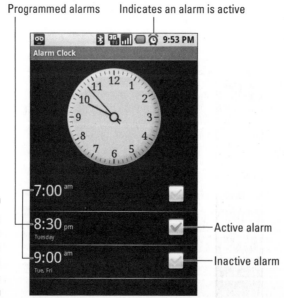

Figure 14-1:
The Alarm
Clock's
home
screen.

Active alarm

Inactive alarm

When you select the box to activate a particular alarm, the Alarm Clock application briefly shows a message letting you know how many days, hours, and minutes remain until that alarm is set to go off (convenient for keeping an eye on how much sleep you're able to get before morning). When any alarm is currently active, you also see an icon of a clock in the status bar to the left of the time (refer to Figure 14-1).

If you have more alarms than can fit on the screen, a scroll bar appears on the right side of the screen. Just scroll through the list with your finger (or the trackball) to see all the alarms.

Oh, and in case you feel compelled to get rid of that big clock at the top so that you see nothing but your list of alarms, just push the Menu button and select Hide clock. (You can always bring the clock back by repeating the process and selecting Show clock.)

Setting a new alarm

To set a new alarm, follow these steps:

1. **Push the Menu button.**

2. **Select the Add alarm menu item.**

   The Set alarm screen appears, as shown in Figure 14-2. At this point, the new alarm has already been created and, by default, has been set to 8 a.m. If you were to push the Back button, you'd see it in your list of alarms. By default, though, it's not active — and odds are good that you want to set it for a time other than 8 a.m. anyway! We'll fix that next.

Figure 14-2:
Creating a
new alarm.

3. **To make this alarm active, touch the check box in the first row, labeled Alarm.**

   You should now see a check mark in the check box.

4. **Touch the second row, Time, to change the time for which the alarm is set.**

   A window appears for editing the time. Touch the plus and minus buttons adjacent to the hour and minute to increment and decrement them, respectively. Touch the box to the right of the time to toggle between AM and PM.

5. **Touch OK in the time window to confirm the new alarm time.**

6. **To change the tone used to signal the alarm:**

   a. **Touch Ringtone.**

      A list of available alarm tones appears.

   b. **Touch a row to select it, which also previews the tone over the loudspeaker.**

   c. **Once you've found a tone you like, touch OK.**

      Or touch Cancel to revert to the previous selection.

7. **If you want the phone to additionally vibrate when the alarm goes off, make sure that the Vibrate check box is selected. Otherwise, deselect it.**

8. **To schedule this as a recurring alarm:**

   a. **Touch Repeat.**

   b. **Select one or more days of the week on which this alarm should repeat.**

   c. **Touch OK to confirm you choices or Cancel to discard them.**

   If you don't select any of the days, the alarm will go off only once — the next time the scheduled time is reached. However, the alarm remains in your list of alarms so you can reenable it simply by going to the Alarm Clock's home screen and selecting the box next to it.

When you've finished tweaking this alarm's options, you can push Back to return to the list of alarms or simply exit the Alarm Clock application altogether by pushing Home. Your alarm has been set.

Editing and deleting alarms

To edit an existing alarm, touch its row in the Alarm Clock's list of alarms (refer to Figure 14-1). You'll be taken to the same screen you saw when creating the alarm, and all of the same options are available: You can modify the time, the repeat days, the alarm tone, and so on.

To delete an alarm, enter its editing screen. From there, push the Menu button and select the Delete alarm item. No confirmation window asks if you're sure that's what you want to do; your alarm will be immediately deleted, no questions asked.

Calculator

With Google having a reputation as a geeky company, you might think that it'd have bundled Android with a ridiculous over-the-top monster of a graphing scientific calculator, but you'd be wrong. The Calculator application is just about the simplest part of Android you'll come across.

The Basic panel

Start the Calculator by selecting its icon in the Applications tab. You are taken straight to the Basic panel, shown in Figure 14-3.

Think of the Basic panel as nothing more than an ultra-simple four-function calculator, the kind you might find for sale for a couple of dollars at an office supply store. It's good for calculating tips, splitting a bill, or adding the number of times the G1's Maps app has bailed you out of a tight situation, but that's about it. You can enter numbers and commands by touching the buttons on the screen or by typing them if the keyboard is open. To delete a digit or command, touch the Clear button; to calculate your current entry, touch the equals sign (=) or press the Enter key on the keyboard.

Figure 14-3:
The Calculator application's Basic panel.

When you're using the keyboard to operate the Calculator, you can use the key combinations of Alt+I for subtraction, Alt+. for division, and Alt+8 for multiplication.

The Advanced panel

From time to time, you might find the need for an occasional pi, cosine, or factorial. When that happens, the Advanced panel, shown in Figure 14-4, is there to bail you out. To switch to this extra-fancy view, push the Menu button and select Advanced panel.

The first thing you'll notice here is that your basic buttons — important things such as digits and the equals sign — have all gone away. That's because the Advanced panel is just a few extra functions that augment the Basic panel to turn it into a bare-bones scientific calculator. If you want to work by touching buttons on the screen, you'll need to keep switching between the two panels to perform tasks.

A better option when using the Advanced panel is to open the keyboard, which allows you to enter basic commands using the keyboard alone while scientific functions are always available on the screen.

Figure 14-4:
The Calculator's Advanced panel.

History

You can browse a history of commands you've entered into the Calculator by scrolling the trackball up and down in either the Basic or Advanced panel. Each upward flick of the trackball will move to the next older entry. Once you've called up an entry from your history, you can recalculate it by touching the equals sign again or first modify it using the keyboard and the Clear button.

To clear your history (you wouldn't want your rivals finding out all of your earth-shattering calculations, after all), push the Menu button and select Clear history.

YouTube

Finally, it's time to kick back and watch some videos! YouTube on the PC has been a fabulous way to blow off a few minutes (okay, sometimes way too many minutes), and Google has done a commendable job of repackaging the experience for Android and the G1 — in fact, unlike some other phones with YouTube capability, the G1 gives you access to YouTube's entire library of videos (I hope you have some free time.) In this section, I walk through the process of browsing and searching for enough video entertainment to ruin your productivity for the rest of the day — or at least until your G1's battery runs out of juice.

You don't have to feel *too* guilty about wasting some quality time on YouTube videos here, because the G1 will make sure that you don't miss calls. When a call comes through, the YouTube application is interrupted and you see the caller information screen as usual.

The Home page

The first screen you see when you open YouTube from the G1's Applications tab is what Google calls the Home page (see Figure 14-5). At the top of the screen, a horizontal bar of video stills appears — these are YouTube's currently featured videos. By touching and holding on this bar and then moving your finger left and right, the bar will scroll. The video still that is currently in the center is the highlighted video, and below it you'll see the video's duration, user rating on a scale of one to five stars, the name of the video, and its description.

YouTube's currently featured videos

Highlighted video   Highlighted video info

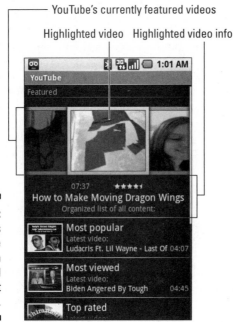

Figure 14-5:
YouTube's
Home
screen with
featured
videos at
the top.

Below the video stills, a list of "most" and "top" categories appears: Most popular, Most viewed, Top rated, Most recent, and Most discussed. Touching any of these will take you to a list of videos currently in the category.

Browsing videos by category

By pushing Menu from the YouTube Home page (or anywhere else in the YouTube application) and selecting Categories, you'll be taken to a list of video categories — a convenient way to just browse around for neat stuff to watch. Once you select a category, you see a list of videos in that category. These lists are perpetual, meaning there's no end to them — as you scroll to the bottom, the application loads more videos. (Theoretically you could reach the end by exhausting YouTube's supply of videos, but let's just say that isn't likely to happen — especially not on your G1.)

Searching

Searching in YouTube works much the same way as in most of the other Android applications you've used so far:

1. **Push the Menu button.**

2. **Select the Search menu item.**

   You can also press the Search key on the keyboard or simply start typing — YouTube guesses that you're trying to search for something and displays a search bar.

3. **Enter your search term(s) in the text box that appears at the top of the screen.**

   As you type, the YouTube application looks through your search history. If it finds anything that matches what you're typing, it displays it in a list below the search bar. To execute that search immediately, just touch the item in the list.

4. **To run the search, press Enter or touch the Search button.**

If you want to clear your search history so that no suggestions are made as you type in the search bar:

1. **Push the Menu button.**

2. **Select the Settings menu item.**

3. **Touch Clear search history.**

   A confirmation window appears, warning you that your search history will be cleared.

4. **Touch OK.**

Video lists

Whether you've searched by keyword or browsed a category, you'll eventually find your way to a list of videos, as shown in Figure 14-6.

From here, you can see thumbnail stills for every video in the list, their names, snippets of their descriptions, their run times, and their user ratings. By simply touching a row here, you'll start playback (more on that in the next section).

By touching and holding on a video, you'll get a pop-up menu of commands:

- ✔ **View:** This command plays the video and is the same as touching the video's row in the list.

- ✔ **Details:** This option takes you to a Details screen for the video (see Figure 14-7), where you can see its full description, the number of times it's been viewed, the user who uploaded it, the day it was added, and its YouTube URL. You can also get direct access to related videos (or at least videos that YouTube thinks are related) from here.

Figure 14-6:
A list of videos in YouTube.

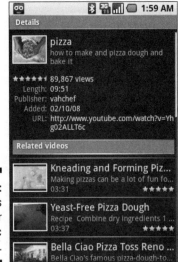

Figure 14-7:
The Details screen for a specific video.

✔ **Add to favorites:** This option adds this video to your Favorites list.

✔ **Share:** The Share command opens Gmail and automatically creates a new e-mail with the direct URL to this video in the body and the phrase "Here's a cool YouTube video" in the Subject line. All you need to do is fill out the To box with a recipient. (Of course, you're welcome to edit the Subject and body fields as well.)

✔ **Comments:** If comments have been left for the video, this menu item appears. Selecting it takes you to a list of comments and the names of the users who left them.

If the video list that you're seeing is in one of the "most" or "top" categories at the bottom of the Home page, pushing the Menu button will reveal a special item: Time categories. Selecting this option allows you to adjust the period of time covered by the list. For example, if you're browsing the Top rated list, Time categories allows you to select whether you're seeing top-rated videos from today alone, all of this week, all of this month, or all time. (The default is just today.)

Watching videos

Once you touch a video in any one of YouTube's many lists, it starts playing. After briefly buffering, the video appears along with controls at the bottom of the screen, as shown in Figure 14-8. From left to right, the three buttons allow you to jump backward by 5 seconds, pause playback (and resume it once it's been paused), and jump forward by 15 seconds. Below those buttons, a bar appears. On the left, the current playback position (in minutes and seconds) is shown. On the right, you see the total duration of the video.

Current playback position

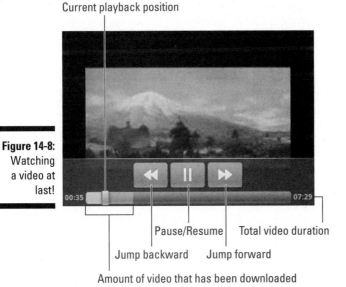

Figure 14-8:
Watching
a video at
last!

Pause/Resume   Total video duration

Jump backward   Jump forward

Amount of video that has been downloaded

A gray box on the bar indicates the current playback position, and the yellow area represents the total amount of the video that has been downloaded to the phone. (Bear in mind that downloading complete videos to the G1 takes a little longer than you might be used to with your PC.) You can move the gray box left and right with your finger to set the current playback position.

When a moment passes without the screen being touched, the controls at the bottom disappear. Just touch the screen to bring them back.

Because Wi-Fi networks offer considerably higher bandwidth than cellular networks do, YouTube automatically shows you a much lower-quality version of videos when you're not connected to Wi-Fi. If you plan on spending more than a minute or two watching videos and you happen to be within range of a Wi-Fi network that you can connect to, it's worth your while to get the better quality picture.

Pushing Menu while watching a video allows you to do many of the same things you can do from the pop-up menu you get when touching and holding on a video in a list — you can add the current video to your Favorites, see its Details screen, or share it with a friend via e-mail.

When you have finished watching the video, just push the Back button to return to the last list of videos you were looking at. You can also push Menu and select Home page to return back to YouTube's topmost screen.

Favorites

If you push Menu from many screens in the YouTube application, you see an item labeled Favorites. Touching this item leads you to the Favorites screen, which is simply a special list of videos that you've added to your Favorites over time. The Favorites list functions the same as any other list, allowing you to play back videos by touching their rows, get video details, or share their URLs via e-mail. To remove a video from your Favorites list, touch and hold on its row and select Remove from favorites from the pop-up menu that appears.

Chapter 15

The Settings Application

In This Chapter

▶ Connecting to Wi-Fi networks, configuring Bluetooth, and controlling cellular networks

▶ Restricting numbers your phone can call and setting other phone-specific options

▶ Controlling your phone's sound and display

▶ Determining whether to synchronize Gmail, Calendar, and Contacts

▶ Deciding how (and if) the G1 can try to determine your location

▶ Locking your G1 from intruders and setting other security features

▶ Managing and deleting applications

▶ Checking your G1's internal and external memory

▶ Setting the date and time

▶ Using keyboard time-savers

▶ Getting info about your phone

*H*ow bright do you want your screen? What wireless networks should you connect to? What date format would you like to use? A phone as complete and powerful as the G1 doesn't become truly usable until you personalize it and make it your own — and part of that process involves configuring its many, many options.

Although many G1 applications have a Settings screen for application-specific options, Android itself also features a catchall Settings application that manages a plethora of system-wide options. Some items that you find in the Settings app affect all applications equally, others modify the way the G1's hardware operates, and still others help you keep tabs on how your phone is operating.

There's lots of good stuff in the Settings application — so much, in fact, that it can be a little intimidating at first glance. Not to worry, though, because in this chapter we walk through every last setting in the Settings application in glorious detail. The sections are organized in the same way as they appear in the application itself, so if you find yourself needing to quickly reference a particular setting down the road, just jump right to its section in this chapter!

Opening the Settings Application

Occasionally, you might find that certain applications on the G1 will automatically open the Settings application for you. A good example of this occurs in Amazon MP3 (see Chapter 16) when you attempt to download purchased tracks without Wi-Fi enabled and connected. Amazon MP3 asks if you'd like to connect now, and if you select Yes, you are whisked straight into the Wi-Fi settings screen of the Settings application. How very convenient!

Usually, though, you open Settings through more conventional means. You can select its icon from the Applications tab, or you can push Menu from the Home screen and select the Settings menu item.

The first screen you see is a list of every category of setting that's available to you, as shown in Figure 15-1. Yes, that's right, I said *category* — Android has so many individual settings that they need to be broken into bite-sized subsections — and what's more, you can scroll down to see even more! To enter a category, simply touch it or select it using the trackball.

And now without further ado, let's jump right into the first category: Wireless controls.

Figure 15-1:
The main
Settings
screen,
showing
many of the
categories
available.

Wireless Controls

As its name implies, the Wireless controls category (see Figure 15-2) covers everything wireless about your phone: Wi-Fi, Bluetooth, and your carrier's cellular network.

Figure 15-2:
The
Wireless
controls
screen.

Here's a look at the settings:

✔ **Wi-Fi:** You can think of this option as the power switch for the G1's Wi-Fi radio. Touching the check box to select it turns on Wi-Fi, but you still need to configure Wi-Fi to connect to specific wireless networks before you can use it. To conserve battery power, you can deselect the box, which powers down the Wi-Fi radio and causes your phone to use your carrier's wireless network exclusively for sending and receiving data.

✔ **Wi-Fi settings:** This option takes you to a separate screen that lets you configure which Wi-Fi networks your G1 will use. I cover this screen in the next section.

✔ **Bluetooth:** Just as the G1's Wi-Fi radio gets a power switch, its Bluetooth radio gets one, too — and this option is it. Selecting this box enables Bluetooth reception.

This check box must be selected for your Bluetooth headset to work with the G1. (You can safely deselect the check box when the headset is not in use, however — the headset will spring back to life as soon as you select it again.)

✔ **Bluetooth settings:** This option opens a new screen where you can set specific Bluetooth-related options and pair Bluetooth devices. I go through this screen in depth in a moment, in the "Bluetooth settings" section.

- **Airplane mode:** By selecting this check box, every single radio inside the G1 — Wi-Fi, Bluetooth, and cellular alike — is turned off in one fell swoop. It's called Airplane mode because turning this option on meets airlines' requirements that electronic devices not be transmitting any signals during flight, which should allow you to continue to use your phone to listen to music or run applications as you fly the friendly skies. (Of course, this also means you won't be able to check your e-mail, make phone calls, or browse the Web, but at least you won't have to turn off the G1 completely.)

- **Mobile networks:** Selecting this option brings you to a new screen that controls how the phone behaves when connected to your carrier's (or another carrier's) cellular network. We take a closer look at this screen shortly, in the "Mobile networks" section.

Wi-Fi settings screen

Touching the Wi-Fi settings item in the Wireless Controls screen takes you to a new screen divided into two sections, Wi-Fi settings and Wi-Fi networks. Under Wi-Fi settings, you have the following two options:

- **Wi-Fi:** This check box has the same effect as the Wi-Fi check box on the main Wireless controls screen; select it to turn on the G1's Wi-Fi radio.

- **Network notification:** If you select this check box, you receive a notification in the G1's status bar whenever it detects that you're within range of an open (that is, unencrypted) and available Wi-Fi network. By pulling down the notifications screen and touching the Wi-Fi notification, you are brought back to the Wi-Fi settings screen, where you can choose to configure the new network.

  This is a handy option when you're in an area without high-speed 3G cellular coverage or you're roaming on a carrier other than your own where consuming cellular data might cost you an arm and a leg. Just select the Network notification check box, wander around, and let your phone find an available Wi-Fi network for you.

In the bottom portion of the screen, the G1 displays all the Wi-Fi networks that it can currently find around you. The phone periodically refreshes the display, but you can manually force it to scan for new networks by pushing the Menu button and selecting Scan.

To connect to a Wi-Fi network, just touch its name in the list. If the network is encrypted for security, you are prompted for a key or a password depending on the type of security in use (see Figure 15-3). After you've entered the key or password, touch OK, and the G1 attempts to connect to the network and use it to transfer data.

Figure 15-3:
Enter your wireless password to continue connecting to this Wi-Fi network.

Some Wi-Fi networks are configured so that they can't be "seen" by devices like the G1. That's not a problem, though, if you know the name of the stealth network. Scroll to the very bottom of the list of available networks and touch the item labeled Add a Wi-Fi network. You see a window where you can enter the network's name (also known as its *SSID*) and tell Android the type of security the network uses. If you choose a security type other than None, you are asked to provide the security key or password, so be sure to have that handy. When you've entered all the requested information, touch Save.

Once you touch a network name in the list (or manually enter it) and configure its security settings, the network remains in the list permanently; it's a "remembered network," and you won't have to configure its security settings again the next time you're within range of it. To make the G1 forget a remembered network, touch and hold on it in the list of networks and then touch Forget network (or, to continue remembering it but change the security password used to connect, touch Change password).

When Wi-Fi is turned on and you are within range of a remembered network, the G1 uses it automatically because Wi-Fi networks are typically much faster than even the fastest cellular networks for transferring data. In real terms, that equates to faster Web browsing, faster e-mail, and most importantly, higher-quality YouTube videos.

If the Wi-Fi network to which you're connecting requires a static IP (most don't), push Menu and select Advanced. Select the Use static IP check box and touch the fields below it and enter network configuration parameters as necessary.

If you're connecting to a public Wi-Fi network, such as in a coffee shop, library, or hotel, you often need to first open your browser and attempt to open any Web page before you can browse the Web, check your e-mail, or do anything else that uses the Internet. When you try to open that first Web page, you are usually redirected to an informational page about the network you're connecting to; the page includes pricing if the network isn't free. If you have an account for the network, this page is where you log in; otherwise, you might have to enter credit card information, a room number, or simply agree to some terms and conditions. After you've taken care of that stuff, you can go back to using the Internet on your G1 as usual.

Bluetooth settings screen

In the Bluetooth settings screen, you configure how your G1's Bluetooth transceiver operates and what devices it can connect to. The top part of the screen displays the Bluetooth settings section, which contains three items:

- ✔ **Bluetooth:** This item does the same thing as the Bluetooth check box in the Wireless controls screen; selecting it enables the Bluetooth radio.

- ✔ **Discoverable:** When you select this box, your G1 can be "discovered" by Bluetooth devices that are actively scanning for other Bluetooth devices. You must enable this option when you are attempting to pair the phone to a computer or another mobile device, but not when you are pairing with a headset (which I cover in Chapter 4).

  Because allowing your G1 to be discovered by anyone who feels like scanning for Bluetooth equipment is a minor security concern (and an annoyance, since others can repeatedly request to pair with you), the option automatically turns off two minutes after it has been turned on.

- ✔ **Device name:** When the G1 is set to be discoverable, the Device name is the name that those actively scanning see in their list of available devices. Touch the row to change it.

Toward the bottom of the screen are any discoverable Bluetooth devices the G1 has found during its automatic scan. (You get bonus points if you find someone else's G1 in this list!) You can perform a fresh scan by pushing Menu and selecting Scan for devices. To pair with a device, just touch its name in the list; if a passcode (also known as a PIN) is required to complete the pairing, you are prompted for it in a window that appears.

Once a device is paired, it remains in the Bluetooth devices list permanently, and your G1 will automatically try to connect to it whenever it detects that the device is on and nearby. (Most importantly, this means your headset is automatically ready to use the moment it's turned on as long as you previously paired it.) To unpair a device in this list, touch and hold on it, and select Unpair in the pop-up menu that appears.

At the time of this writing, the G1 supports only the Hands-Free and Headset profiles of the Bluetooth standard, which basically means that headsets and external speakerphones (such as hands-free devices for your car) are the only gadgets worth pairing at the moment. You'll still be able to pair to other devices such as computers, but until (and if) Google and HTC decide to provide updated software for the G1 that supports other profiles, performing these pairings won't do you much good.

Mobile networks settings screen

The Mobile networks settings screen gives you a handful of options for controlling how and when the G1 makes use of the cellular networks around it. Let's take a look at the options here:

✔ **Data roaming:** When this option is selected, the G1 continues to send and receive data even when it is connected to a network other than T-Mobile. Be warned — this can get very expensive when you're outside the United States, so unless you're ready and willing to rack up some serious charges in the name of being able to check your e-mail, you're probably best off leaving this option disabled.

✔ **Use only 2G networks:** Although the G1 can connect to the newer, faster 3G networks that T-Mobile now has available in many cities, the speed comes with a tradeoff: battery life. To eke out every last drop of juice from your battery, select this box, which forces the G1 to use 2G networks at all times, even when a 3G network is available. (This can be an especially advantageous option when you know that you are connected to a Wi-Fi network, which is faster than 3G anyway.)

When your G1 is searching for a 3G network and can't latch onto one, that's particularly draining on the battery. If you're in an area where there's no 3G coverage (you'll know because you never see the 3G icon in the G1's status bar) or the phone periodically switches between EDGE and 3G mode, you can get some extra useful life out of your battery charge by making sure you select the Use only 2G networks check box.

✔ **Operator selection:** Under most circumstances, your G1 will automatically select the best cellular network it can find and connect to it. If you want to manually override that selection, you can touch this option, which presents a list of all networks the G1 can currently find; touch a network name in the list to connect to it. Note that you'll be able to connect only to networks with which T-Mobile has a roaming agreement in place; so, for example, you can't typically connect to AT&T in the United States, even if you can see its network in this list.

✔ **Access point names:** An access point name (APN) is simply a handful of important configuration settings that tells your phone how to transmit data on a particular cellular network. To send and receive data, the G1 — like any cellphone — requires a valid APN for the network to which you're currently connected. Fortunately, the G1 comes installed with all the APNs required to connect to any T-Mobile network anywhere in the world where it provides service, so you won't usually have to worry about this screen.

If you travel overseas to an area not serviced by T-Mobile and use a local carrier's SIM card in the G1, and you find that you do need to create a new APN, you typically do so with the help of documentation or a carrier's customer service agent. From the Access point names screen, push the Menu button, select New APN, and fill out the fields on the screen according to the instructions you're given. (Note that you may not need to fill out every field depending on the network that you're connecting to.)

Before you can install a local carrier's SIM card in the G1, you'll need T-Mobile's assistance to unlock the phone, which allows non-T-Mobile SIM cards to be used. Just give customer service a call to get help with unlocking.

Call Settings Screen

With all the whiz-bang applications that Android features, it's easy to lose sight of the fact that the G1 is still a phone at heart. But indeed it is, and the Call settings screen, shown in Figure 15-4, helps you configure many of its phone-centric options.

In this screen you have:

✔ **Fixed Dialing Number:** This feature lets you restrict the phone numbers your phone can call. It's a convenient option when you're lending your phone to someone and you'd like to keep the random international calls to Timbuktu to a minimum; it's also good for a phone that you're giving to a child who you only want to be able to call you. I discuss Fixed Dialing Number (FDN) in detail in the next section.

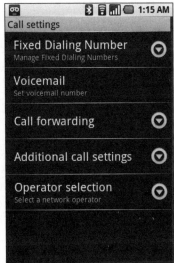

Figure 15-4:
The Call
settings
screen.

- ✔ **Voicemail:** Touching this option allows you to change the number that is called when you dial voice mail (usually by holding down the 1 key in the Dialer). Unless you have a specific reason to change this phone number (for example, T-Mobile has actually changed the number they use for their voice mail system and told you to update your phone accordingly), you shouldn't have to worry about it since your SIM card has your correct voice mail number built-in.

- ✔ **Call forwarding:** Touching this option takes you to another screen where you can set up call forwarding for a variety of situations. By default, your phone is set up to forward to your voice mail number when the phone is busy, is unreachable (meaning the outside cellular coverage or turned off), or goes unanswered. The forwarding number for each of these situations can be changed or disabled independently.

  Additionally, you can use this screen to set up the G1 to always forward directly to another number, regardless of the situation. To do so, touch the topmost item, Always forward, and then type the forwarding phone number (or touch the button to the right of the text box to select a number from your contacts list).

- ✔ **Additional call settings:** Selecting this option opens a screen where you can configure call waiting and choose whether your phone number is displayed to people you call. (You may know this feature better as caller ID). If call waiting is enabled, you are notified when someone calls you while you're already on a call. If call waiting is disabled, the call forwarding rule for when your phone is busy applies. By default, that rule sends callers directly to your voice mail.

✔ **Operator selection:** This option works the same as the Operator selection screen in the Mobile network settings screen within the Wireless controls screen (whew), allowing you to choose a different cellular network than the one the G1 has selected for you by default.

If you want to use the Fixed Dialing Number (FDN) feature to restrict what numbers can be called from your phone, select the Fixed Dialing Number item from the list. You are taken to the FDN screen, shown in Figure 15-5.

To turn Fixed Dialing Number on or off, you need to know a special passcode called PIN2 that is programmed into your SIM card. For T-Mobile, the default PIN2 is usually 5678. If that doesn't work and you can't figure out the correct PIN2, give customer service a call.

Before enabling FDN, you want to set up the list of phone numbers that can be called. Now that you're in the FDN screen, do the following:

1. **Touch Manage FDN list.**

   You see your list of FDN phone numbers. If you haven't set up this feature in the past, the list is probably empty.

2. **Push Menu and select Add contact.**

3. **Enter a name and phone number for this contact.**

4. **Touch Save.**

   Now, repeat Steps 2 through 4 for each contact you want to be able to call when the FDN feature is turned on.

5. **When all contacts have been entered, push Back.**

   You return to the main FDN screen.

6. **Touch FDN disabled.**

   You are prompted for your PIN2.

7. **Enter your PIN2 and touch OK.**

   As I mentioned, the PIN2 is probably 5678 unless you changed it. If your PIN2 was correct, the FDN disabled item changes to FDN enabled. At this point, only contacts you've specified can be called from the phone.

To disable FDN in the future, just touch FDN enabled, enter your PIN2, and touch OK. You can change your PIN2 (since the default of 5678 isn't particularly secure) by touching the Change PIN2 item on this screen.

Figure 15-5:
Configuring
the Fixed
Dialing
Number
feature.

Sound and Display Settings Screen

News flash: Your G1 has a huge, bright, beautiful screen and a loud, clear speaker. Then again, you probably already knew that, but what you might *not* have known is how to get those lovely components under your complete control. It turns out that the Settings application's Sound and Display settings screen, shown in Figure 15-6, is your answer.

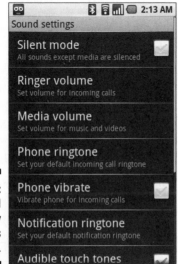

Figure 15-6:
The Sound
and Display
settings
screen.

Here's what is available in the Sound settings section, at the top of the screen:

- ✔ **Silent mode:** By selecting this check box, your phone won't make a peep (unless you play music or videos). This option is the same as holding down the End button and selecting Silent mode from the pop-up menu that appears.

- ✔ **Ringer volume:** Touching this option displays a slider that you drag left and right with your finger to set the ringer's volume. This action performs the same function as pushing the G1's volume buttons up and down (when you're not in a media application such as Music or YouTube, anyway).

- ✔ **Media volume:** This option displays a volume slider just like the Ringer volume option does, but this slider determine the loudness of your music and videos. Using this option does the same thing as pushing the volume buttons when you're in a media application.

- ✔ **Phone ringtone:** Touching this option calls up a window that lets you choose your G1's default ringtone for incoming calls. Touching an item in the list plays it. When you hear a ringtone you like, touch OK to confirm your choice. To back out without changing, touch Cancel.

Although the Phone ringtone option chooses the default ringtone, you can override your selection for individual contacts by editing the ringtone setting within the contact's details from the Contacts application. Check out Chapter 8 to see how to do that.

- ✔ **Phone vibrate:** When you select this option, the phone vibrates when you receive a call. This feature is handy when the G1 is doing time in your pocket and you may not hear a ring.

- ✔ **Notification ringtone:** This option lets you choose the default sound you hear when you receive a notification in the G1's status bar. (Keep in mind that this sound can be overridden by individual applications that post their own notifications.)

- ✔ **Audible touch tones:** When you select this option, you actually hear the touch tone sound when you press a number on the Dialer's keypad — just like that old AT&T Slimline of yours. Nostalgia!

- ✔ **Audible selection:** By selecting this box, you hear a brief tone each time you touch something on the screen. I find it annoying. Unless you have a specific reason to use this option, you may want to leave it deselected. (Your coworkers at the meeting table will thank you.)

Below these options, you see the Display settings section, which contains two items:

- ✔ **Brightness:** Touch this option to display a slider that you can drag with your finger to control the brightness of the screen's backlight.

TIP

You can squeeze out a few precious minutes of extra battery life by keeping the screen's brightness at a lower setting.

✔ **Screen timeout:** This option sets the amount of time that elapses without any input from you before the screen turns off to conserve power. If you find yourself frequently putting your phone down and reading the display without touching it, you may want to choose a higher setting (say, a minute or two).

Data Synchronization Screen

As you've experienced from the moment you first powered up your little hunk of Google circuitry, Android effortlessly and automatically synchronizes your most important data — your Gmail, Calendar, and Contacts — with Google's servers on an ongoing basis. That way, you can always count on being able to access your latest and greatest data from either your PC (through Google's Web site) or your G1.

Android's data synchronization capabilities are powerful, but you may want to turn them off at times. Constantly keeping your G1 up-to-date requires data consumption and power, so if you're in a situation where you need maximum battery endurance, you might want to temporarily prevent the G1 from phoning home for updated information. You can make this happen from the Data synchronization screen, which is shown in Figure 15-7.

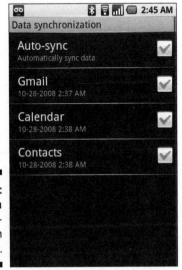

Figure 15-7:
The Data synchronization screen.

As you can see, all the check boxes are selected by default, which means Gmail, Calendar, and Contacts are updated automatically and continuously.

By deselecting the first box, Auto-sync, you prevent all three Google services from automatically synchronizing and you have to initiate synchronizations manually until you reenable the Auto-sync setting.

To force a synchronization immediately, push the Menu button from this screen and select Sync now. You can use this command whether Auto-sync is enabled or disabled.

Below each of the three Google services in the list, you see the date and time they were last synchronized. You can prevent individual services from synchronizing (useful if you're using Gmail and Contacts but not Calendar, for example) by deselecting them.

Location and Security Screen

The screen shown in Figure 15-8 lumps two very important concepts — location services (such as Maps) and phone security features (such as locking) — into a single screen. Why are they lumped, you ask? The logic is that your current location *is* a matter of security, in a way. Just as you may not want certain people to know where you are at every waking moment, there may be times when you don't want applications on your G1 to be able to find out where you are, either.

Figure 15-8:
The location and security screen.

My Location sources section

At the top, the My Location sources section is where you decide how the G1 is allowed to try to determine where you're located:

- ✔ **Use wireless networks:** When you select this check box, the G1 can take an educated guess at your location by using what it knows about the location of cellular towers and Wi-Fi networks. Even if GPS is enabled, it's a good idea to keep this option enabled (especially when using features such as My Location in the Maps application) because GPS doesn't work well indoors or in areas where you don't have a great view of the sky.

- ✔ **Enable GPS satellites:** As I mention in Chapter 7, GPS is far more accurate than using wireless networks alone to determine your location, but that accuracy comes at a cost — battery life. You can help keep your G1 juiced for a bit longer by turning off this option. Unless you're using the G1 to help you with turn-by-turn navigation while driving, you may find that the G1 is capable of estimating your position well enough without this option.

Screen unlock pattern section

The next section of the screen, Screen unlock pattern, lets you do something unusual (and very cool) that I can guarantee you haven't seen on any other cellphone. After a typical phone's screen turns off to conserve power, many models can be configured to require a numeric passcode once they are powered back on. This small, quick security measure prevents anyone pilfering your handset from seeing your data and placing calls.

Android allows you to lock your phone to make it difficult for would-be snoopers, too, but it takes the concept to another level. When you set an unlock pattern on your G1, you physically trace a pattern on the screen with your finger, and the phone unlocks only if you trace the correct pattern. It works extraordinarily well, and you might find that you can unlock your phone faster with a pattern than by typing a passcode.

To set an unlock pattern:

1. **Touch Select unlock pattern.**

   You see a screen that describes the process of setting and using an unlock pattern. To back out without creating a pattern, just touch Cancel.

2. **Touch Next.**

   A brief video demonstrates how to trace your finger to produce a pattern. You need to run your finger across at least four dots to create a valid pattern (any fewer than four dots are too easy for someone to guess).

3. **Touch Next again.**

   Now you're ready to draw your pattern (see Figure 15-9). Touch and hold your finger on one of the dots; then drag to connect adjacent dots of your choosing until you have a pattern that you're happy with (see Figure 15-10). As you drag, a gray line shows the pattern you've created thus far. Green arrows indicate the direction of the pattern.

   If you're unhappy with the pattern you've created, touch the Retry button in the lower left of the screen; the pattern will be cleared and you can start over. Be sure to choose a pattern that you can remember — you need to draw this pattern every time you go to use your phone!

4. **When you've drawn a pattern you like, touch Continue.**

5. **Draw the same pattern again to confirm your choice.**

6. **Touch Continue.**

At this point, your pattern is set and ready to use. If the Require pattern check box (in the Screen unlock pattern area) wasn't selected before, it is automatically selected for you at this point. This check box is the on/off switch for pattern locking; when it's selected, the pattern is required any time the screen is turned on after it has been off.

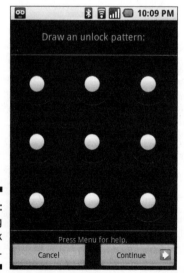

Figure 15-9:
Setting
an unlock
pattern.

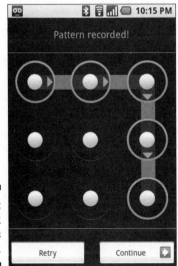

Figure 15-10:
An unlock pattern has been drawn.

Below the Require pattern option is the Use visible pattern check box. When you select this option, the pattern you draw on the screen to unlock the phone is visible — that is, as you draw from dot to dot, a line appears. Being able to see what you're drawing makes unlocking the phone a little bit easier, but the tradeoff is that it's less secure because the people around you can see the pattern you're drawing.

So what does it look like when the time comes to unlock your phone after it's been sitting idle for a while? By pushing any button on the front of the phone except Menu, the G1 springs to life and shows you its standard standby screen (see Figure 15-11). If you push Menu, you go straight to the Draw pattern screen instead.

If you haven't already done so, you can push Menu to see the Draw pattern screen. Simply drag your finger from dot to dot to draw the pattern you've set. As soon as you complete the pattern, the phone unlocks and you see whatever screen was active before the G1 was locked.

Because the screen unlock pattern feature prevents users of your G1 from dialing phone numbers while the phone is locked, a special "escape hatch" allows anyone to make a call to 911 (or other emergency services) without unlocking the phone. To get to this feature, touch the Emergency call button at the bottom of the Draw pattern screen — you go straight to a Dialer keypad where you can enter the number you need to dial.

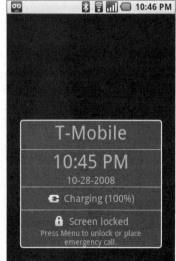

Figure 15-11:
The G1's
standby
screen.

If you're really concerned about making sure that there's no reasonable way for intruders to find their way into your phone, be careful — and make sure your hands are clean! Grubby fingers could leave a trail of grease on the screen that makes your unlock pattern visible to the naked eye.

SIM card lock section

In addition to your G1's pattern lock, the SIM card inside your G1 features its own type of lock. When you enable this lock, you are prompted for a four-digit number (known as your PIN code) whenever the phone is turned on.

To enable the SIM card lock on the G1, touch the SIM card lock item (refer to Figure 15-8). You are taken to a new screen, SIM card lock settings, where you can enable and disable the lock and change your PIN code.

Unless you've previously changed the PIN code on your T-Mobile SIM card, the code should be 1234. You need to know your PIN code whenever selecting or deselecting the Lock SIM card check box in this screen. I strongly recommend changing your PIN code if you intend to use this feature; 1234 isn't the most unguessable combination of numbers in the world.

Passwords section

The Passwords section contains only one option: Visible passwords. When this option is selected, the last letter is visible as you type passwords anywhere in Android (such as the IM application when you're entering the username and password for a new account). As soon as you type another letter, the previous letter changes to a dot so that the entered password cannot be seen — this makes it a little easier to verify that you're typing the password correctly. When you deselect this option, characters in passwords turn into dots as you type them.

Applications Settings Screen

The G1 isn't just about the applications it includes out of the box. The phone is also about everything you add to it. Just like with your PC, you can install software to get the features, entertainment, and functions that are important to you. The Applications settings screen, shown in Figure 15-12, is your home base for finding out what apps are installed, managing them, and getting rid of them if you're so inclined.

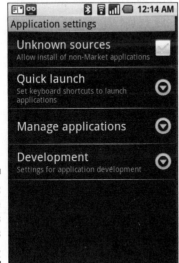

Figure 15-12: The Applications settings screen.

Unknown sources section

The top of the Applications screen, the Unknown sources section, controls where the G1 will let you go to install new applications. By default, the G1 will allow you to install only apps downloaded from Google's official Android Market (see Chapter 18). Software from other sources — the Web, e-mail attachments, and so on — is blocked from installation.

The Internet is a jungle, and just like on your PC, grabbing Android applications from sources you're unfamiliar with is risky because you have no way of knowing whether the apps you download will damage your G1 or your data. Keeping the Unknown sources check box deselected keeps you safer by limiting you to using applications that Google is aware of, but the tradeoff is that you won't have access to as many applications. Be careful!

Quick launch section

Did you know that you can launch the Browser no matter where you are in your G1 by pressing the Search key and the B key at the same time? What's even better is that you can choose your own quick launch key combinations.

Touching the Quick launch screen item takes you to a new screen displaying the letters *A–Z* and the numbers 0–9 listed on the right side. These letters and numbers represent keys that make up part of each quick launch key combination; the other part of the combination is the Search key. On the left is the name of the application assigned to the combination. For example, you see that Search+B opens the Browser, Search+C opens Contacts, Search+G opens Gmail, and so on.

By touching any row in the list, you can assign (or reassign) that letter's quick launch combination to an application on the G1. For example, say you want to assign Search+V to Voice Dialer. Here's how you'd go about it:

1. **Touch Quick launch.**

   The list of all possible key combinations appears.

2. **Scroll down in the list to the letter V.**

   By default, no application is assigned to Search+V, so you won't have to overwrite a quick launch that's already configured to create this one.

3. **Touch Assign application to the left of the letter V in the gray box.**

   You see a list of all installed applications.

4. **Scroll down the list of applications and choose Voice Dialer.**

You're all set — you can now press Search+V no matter where you happen to be in the G1's many screens and be taken immediately to the Voice Dialer. To clear the Voice Dialer quick launch, return to the Quick launch screen, find the letter V in the list, and touch and hold on it. After a moment, you are asked if you want to clear the shortcut; just touch OK to proceed.

Manage applications section

If you get a kick out of experimenting with new apps for your G1, trying out pretty much every new package that hits the Android Market in the process, that's great — we're kindred spirits. Like me, you'll become best friends with the Manage applications screen, where you can see at a glance what's installed, how much space each app is taking, and (in the case of apps you've installed, anyway) remove the ones you don't want.

Touching the Manage applications item takes you to a list of all applications on your phone. The list is made up of a combination of built-in Android apps (such as Alarm Clock and Browser), system programs you don't normally see and have no control over (such as Package Installer), and applications you've installed. To the right side of this list, you see the amount of memory each application takes up in your G1.

The default sort order for this list is alphabetical, but sometimes that's not the most helpful way to look at the list. If you want to see the largest applications (also known as "the worst offenders" if your goal is to conserve as much storage space as possible), push the Menu button and select Sort by size (descending).

Now, let's look at a single application to see what we can find out about it. Touch an application's row in the list and you are brought to the application's details screen. The one for Amazon MP3 looks like Figure 15-13.

At the top, you see the application's full name and icon. Below that, the screen is divided into three sections:

✔ **Storage:** In this section, you see the size of the application in kilobytes (KB) or megabytes (MB). Next, the total size is broken into two categories: the size of the application itself and the size of any data that it has stored on the phone. For example, if you were to install a note-taking application, the total size of notes that you had created would be listed in the Data category while the application itself would go under Application.

If the section has a button labeled Uninstall, you can touch it to remove the application. If the application is not removable (usually because it came with your G1 and is built into to its memory), you see a Clear data button instead, which you can use to delete any data that you and the application have created.

Figure 15-13:
The details
for an app
installed on
the G1.

✔ **Launch by default:** If this application has been configured to be the default application for certain actions on the G1, you can reset that here so that the application is no longer automatically launched by touching the Clear defaults button.

For example, when you touch a contact's e-mail address in the Contacts application, a new window appears, giving you the option of opening either the Gmail or the Email application to send a new e-mail. A check box at the bottom of the window allows you to save your selection as your default so that you aren't asked which e-mail application to open the next time. If you've selected Email as your default application but later want to use Gmail instead, no sweat: Touch the Clear defaults button in Settings for the Email application, and the next time you touch a contact's e-mail address, you'll be prompted to select between Gmail and Email again.

✔ **Permissions:** This feature is really neat and unique to Android. With so many developers creating so many applications for the Android platform, it can be tough to decide what's trustworthy enough to install, especially considering how much private and personal data you might have on your phone. Everything on the G1 that this application is allowed to access — such as the Internet and your contacts — is listed here, so you can make a better judgment about whether you're comfortable using the application.

How does this help? You can look for permissions that are out of place. For example, if you've downloaded a one-player solitaire card game and the permissions screen tells you that the application has access to your contacts, that might be a sign that the application will do something that you don't expect it to. Of course, nothing here is a surefire indication that you're dealing with malicious software, but you can use the Permissions feature as a tool to help you make that determination.

Because an application can be granted many kinds of permissions, Android makes things a little easier by highlighting in yellow the most important ones — those that could potentially compromise your data or run up your phone bill. The rest are collapsed into a bar at the bottom of the screen that says Show all; just touch the bar to see the other less critical permissions.

Development section

Returning to the Application settings screen, you see one last item at the bottom titled Development. Touching this takes you to a new screen with two options. Selecting USB debugging allows your G1 to be debugged by connecting it to a PC with a cable. Unless you're an Android application developer, you'll never need to worry about this item and should leave the option deselected. The second option, Stay awake, keeps the screen from turning off while the phone is connected to a PC. (This is a particularly convenient option for developers who also need to stay awake — usually with a combination of coffee and Red Bull.)

SD Card and Internal Phone Storage Screen

The SD card and internal phone storage screen, shown in Figure 15-14, is all about your memory. Well, not *your* memory — your G1's memory. This screen is where you see the G1's total memory and available memory and (cue dramatic music) delete everything on your phone.

The following are the items in the top section, SD card:

✔ **Total space:** The first item shows you the total amount of space available on the microSD memory card in your G1. Because some overhead is involved in formatting a memory card for use with the phone, you won't see the total advertised space here. For example, if you have a 1GB card inserted, you'll see only 968MB available to you.

Figure 15-14:
The SD card
and internal
phone stor-
age screen.

✔ **Available space:** The second item is the total space on your card minus anything that has been stored on it — music, photos, applications, and so on.

Anything that you download, install, or create on your G1 will make use of the memory card, not the G1's internal memory. There's just not enough internal memory to go around!

✔ **Use for USB storage:** When you select this option, the contents of your G1's memory card are made available on your PC whenever you connect it with a USB cable — just like one of those thumb drives you can buy from your local electronics store. The disadvantage of using this option is that only one device can "see" the contents of the card at a time, so when your PC is using it, the G1 can't. I discuss the use of your G1 for USB storage in more depth in Chapter 17.

✔ **Unmount/eject SD card:** To remove the microSD card from your phone, touch this option first.

The Unmount/eject SD card option gives the G1 an opportunity to make sure it's not using the card before you pull it out, preventing potential damage to data.

At the bottom of the screen is another section, Internal phone storage, with two items:

✔ **Available space:** This item is the amount of space available on the G1's internal memory. Because this memory is automatically managed and controlled by the G1, you shouldn't have to worry about it too much.

✔ **Factory data reset:** The final item is the doomsday control — the prover-bial big red button. Touching this item clears all data from your phone and makes it as minty fresh as the day you bought it. Your applications, photos, music, and other items will all be gone, and when the phone turns back on, you'll need to enter your Google account information over again. (If you should accidentally touch this option, don't worry: You need to confirm your choice in the screen that follows. Don't do this unless you really mean it!)

Date and Time Settings Screen

In the past five to ten years, a good percentage of the world's wristwatch wearers have ditched their trusty timepieces. But why? Blame the cellphone. Not only do phones do a perfectly fine job of keeping the date and time, they also have a leg up on your average wristwatch by having their time set auto-matically by your wireless network. Gone are the days of setting the watch twice a year for Daylight Savings (and speaking as a rather lazy individual, thank goodness).

Should you want to set your phone's time or date manually, though, you can do that from the Date and time settings screen (see Figure 15-15).

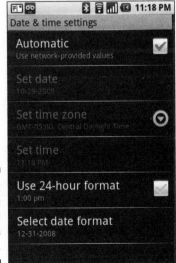

Figure 15-15: The Date and time settings screen.

The Automatic check box is selected by default, which means your G1 is listening to your carrier's cell towers to find the correct date and time. As you move around the world, the time and time zone are automatically set for you as long as you're within network coverage.

If the date or time being sent out by your network is wrong (believe me, it can happen!) or you like to set your time a few minutes slow or fast, you can deselect the Automatic check box. Doing so enables the Set date, Set time zone, and Set time rows, each of which you'll now be able to touch to modify.

Below these rows, the Use 24-hour format check box allows you to configure the G1 to report military-style 24-hour time wherever times are being shown in the software — just select the check box. If the check box is not selected, you see 12-hour time with an AM or PM indicator on the end of it.

Finally, touching Select date format allows you to choose the way that you see dates in Android. By default, you see a two-digit month, a two-digit day, and a four-digit year all separated by dashes. But if you prefer to see a word for the month (such as *Oct* instead of 10) or the year at the beginning of the date, you can choose those options here.

Text Settings Screen

The Text settings screen, shown in Figure 15-16, is where you set options that control how the G1's keyboard behaves when you're using it to edit text fields — in other words, any place throughout Android and your applications where you can enter text.

Here are your options:

- **Auto-replace:** Using a tiny keyboard with your thumbs is an inexact science, no matter how good that keyboard might be! Auto-replace senses common spelling errors when you're typing and automatically fixes them with no fuss. (If you're not looking carefully, you may never even notice you made them to begin with.)

- **Auto-cap:** If you're working with a lot of e-mails — or you're just a stickler for literary perfection in your text messages — Auto-cap saves you some thumbstrokes by automatically capitalizing the first letter in every sentence you type. No Shift key necessary!

✔ **Auto-punctuate:** Like Auto-cap, Auto-punctuate is a minor time-saver that can become a major time-saver if you write a lot of e-mails. By pressing the space key twice in succession, this option assumes that you're trying to end a sentence and start a new one, causing a period to be inserted followed by a space. In other words, you won't need to move your thumb over to the period key nearly as often.

Figure 15-16: The Text settings screen.

About Phone Screen

The About phone screen, shown in Figure 15-17, is a brain dump of information that describes almost everything about your G1's software and hardware: software versions, details on signal strength and battery life, and even a cute scrolling marquee of individuals and companies that have contributed their blood, sweat, and tears to make Android a reality (touch Contributors to see this moving tribute). You may need to come to this screen on occasion to read pieces of information requested by T-Mobile customer service, but otherwise, you probably won't be visiting often.

If you use the MAC filtering feature on your Wi-Fi router at home, you need to add the G1's MAC address to your router to wirelessly connect the G1 to it. To find the MAC address, touch the Status item in the About phone screen, scroll down, and look for the code below the row labeled Wi-Fi MAC address.

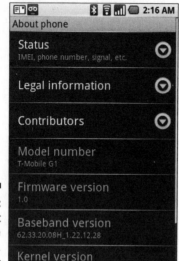

Figure 15-17:
The About
phone
screen.

Part IV
Adding Software and Media

The 5th Wave   By Rich Tennant

"Of course your current cell phone takes pictures, functions as a walkie-talkie, and browses the Internet. But does it shoot Silly String?"

In this part . . .

After you decide to put a G1 in your pocket for the first time, odds are it won't be too long before you're ready to start adding your own music, games, and applications that can make the phone even more fun and useful to you. Chapter 16 shows you how to buy new music right from the comfort of the G1 and Chapter 17 talks about adding music (and pictures, for that matter) that you already own. Chapter 18 takes you into the Android Market, your one-stop shop for cool new applications and games.

Chapter 16

Buying Music with Amazon MP3

*I*f you listen to your music collection on your PC or a portable media player of one sort or another, odds are good that you already have a decent collection of digital music ready to dump into the G1's memory card. But even if you don't, haven't you ever been out on the town, heard a song you liked, and wished you could buy the track (or the whole album) right then and there?

With the G1's Amazon MP3 service, you can. In principle, it's a lot like the iPhone's iTunes Wi-Fi Music Store, but most tracks are even cheaper through Amazon's service — and all files are standard MP3s with no DRM (digital rights management) applied, meaning the files you buy on your G1 can be transferred to your computer and your other media players with no worries that they might be incompatible.

Amazon MP3's library of artists and record labels isn't as extensive as undisputed industry leader iTunes, but it's pretty darned close. (Let's put it this way: Unless your taste in music is *really* off the wall, I'm willing to bet you'll be able to find a track or five that you're interested in.) In this chapter, I run through how to preview and buy tracks through Amazon MP3's service and listen to them almost immediately. For anyone who believes in instant gratification, Amazon MP3 is a dream come true.

Starting Amazon MP3

You might think that Amazon MP3 would be a function of the Music application, but it's not. Amazon MP3 is its own application with its own icon in the Applications tab. Why is that? Aren't you going to be playing all the music you buy from Amazon MP3 using the Music application anyway?

Not necessarily! One of the great things about Android is that you're free to replace the built-in music player with a different application if you so choose. Google and Amazon recognize this, so Amazon MP3 is nothing more than a way to buy MP3 files directly from the comfort of your phone; how (and where) you use those files is entirely up to you.

When you first start Amazon MP3, you are brought to its Home screen, shown in Figure 16-1, where you can browse or search for music to sample and buy.

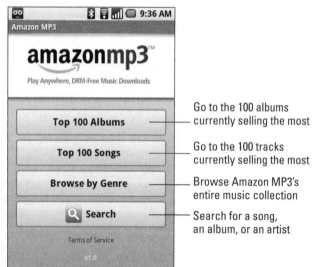

Figure 16-1:
Amazon
MP3's Home
screen.

Go to the 100 albums currently selling the most

Go to the 100 tracks currently selling the most

Browse Amazon MP3's entire music collection

Search for a song, an album, or an artist

Finding Music

When you walk into a record store (do those even exist anymore?), the aisles are organized in a way that helps you find what you're looking for — usually by genre, and by artist within each genre. Amazon MP3 takes some organizational cues from those brick-and-mortar stores of yore, but with a couple of awesome differences. One, Amazon MP3 lets you buy individual tracks from almost any album. Two, your G1 gives you access to millions of tracks instantly — and they're always in stock. Let's take a look at the different ways you can find what you're looking (or browsing) for.

Go with the flow: Finding popular music

If you were doing the Twist in the 60's, the Electric Slide in the 80's, and the Macarena (my goodness) in the 90's, there's a good chance you might be into popular music, and Amazon MP3 makes it easy to get right to it.

On the Home screen, the first button, Top 100 Albums, takes you straight to the 100 albums that are currently selling most briskly on the Amazon MP3 service. The second button, Top 100 Songs, does the same for individual tracks. One of the great things about buying music digitally instead of visiting a record store is that the vast majority of songs found on albums are available individually if you don't care to buy the entire album.

You can find some amazing deals in the Top 100 lists. At the time of this writing, for example, I could buy the Grateful Dead's *Terrapin Station* for just 99 cents (yes, the entire album!).

Browsing the catalog

To browse through Amazon MP3's entire collection of music, select the third button on the Home screen, Browse by Genre. This takes you to a screen that allows you to select a genre of music — Blues, Classic Rock, or Jazz, for example — and optionally drill down to specific subgenres where applicable.

Suppose you want to browse through the available Easy Listening albums. Here's what you'd do:

1. **From Amazon MP3's Home screen, touch Browse by Genre.**

   You are taken to a list of broad genres, such as Alternative Rock and Blues, as shown in Figure 16-2.

2. **Scroll down until you see the Pop genre and touch it (or select it with the trackball).**

   Now you see all the available subgenres within Pop — Adult Alternative, Adult Contemporary, and so on.

3. **Touch the Easy Listening subgenre.**

Figure 16-2:
The Browse
by Genre
screen.

You're now looking at a list of albums within the Easy Listening subgenre, sorted in order of sales popularity. Anxious to buy that new Michael Bublé album? Hang tight for just a moment — I explain the album and track list screens shortly!

If you want, you can also browse genres by individual track popularity instead of album popularity. To do this, return to the Browse by Genre screen (refer to Figure 16-2) and touch the Songs button at the top of the screen. When you select your subgenre, you'll now see a list of songs instead of albums.

Searching for music

If you're like me, when you open Amazon MP3 you want to look for a particular song or artist that you just heard. (Sure enough, searching once again plays a big role in making the G1 great.) Here's how to start a search:

1. **Go to Amazon MP3's Home screen.**

   If you're looking at any screen besides Home in the application, you can get to the Home screen quickly by pushing Menu and selecting the Home menu item.

2. **Touch the on-screen Search button.**

If you already have the keyboard open, you can also press the Search key on the keyboard.

A search bar appears at the top of the screen (see Figure 16-3). Your previous searches appear as a selectable list directly below the text box, so if you want to repeat a search that you recently ran, just touch or select it.

3. **Assuming you haven't selected an old search, start typing the song, album, or artist name in the search bar.**

4. **When you've typed enough of the name to help Amazon drill down to what you're looking for, press the Enter key or touch the on-screen Search button.**

You now see a list of album matches (see Figure 16-4); touch the Songs button at the top of the screen to switch to individual songs that match your search terms.

Previous searches   Search bar

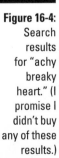

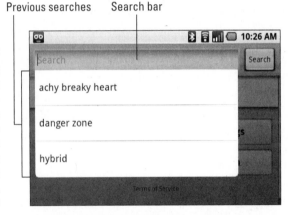

Figure 16-3: The search bar (I promise the search history doesn't indicate my taste in music).

Figure 16-4: Search results for "achy breaky heart." (I promise I didn't buy any of these results.)

Previewing and Buying Music

Once you've found whatever album or track you're looking for — be it by browsing or searching — the logical next step is to plunk down some hard-earned cash for the tracks. Before you do that, though, Amazon MP3 gives you the option of previewing a sample (usually 30 seconds long) of every track you're interested in.

Buying an entire album

Say that you've searched for the album *Gods & Monsters* by Juno Reactor. As I mentioned in the previous section, the first thing you'll see after search-ing is a list of album matches. If you wanted to purchase an entire album "sound unseen" — that is, without first previewing any of its tracks — you can get right down to business from the screen you saw in Figure 16-4:

1. **Touch the price of the album that you're interested in.**

   The price changes to the word BUY, as shown in Figure 16-5. Rather than immediately purchasing the album when you touch the price, you first see this intermediate step so that you don't accidentally brush the screen and buy something you didn't mean to — you're always two clicks away from a purchase.

Figure 16-5:
I'm about to
buy *Gods &
Monsters*
by Juno
Reactor.

2. **Touch BUY.**

   Although you can preview and buy music using T-Mobile's 3G network, you must be connected to a Wi-Fi network to download your purchased tracks. This is because entire music files are large — an album can be 50MB or more — and it's not practical to try to download that much data over a slower cellular connection. If you attempt to buy music when you're not connected to a Wi-Fi network, you can complete the purchase, but you'll see a warning (see Figure 16-6) that you have to wait until you hook up to Wi-Fi before getting your purchased tracks. (I cover the process of connecting to a Wi-Fi network in Chapter 15.) Touch OK here to proceed with the purchase with the understanding that you won't be able to get your music just yet.

3. **Enter your Amazon.com account e-mail address and password.**

   If you want to enable Amazon's 1-Click ordering so that you don't have to enter your e-mail address and password every time you want to make a music purchase, select the Enable 1-click ordering check box by touching it.

   If you don't have an Amazon account, you can set one up quickly and easily by visiting Amazon's Web site at `amazon.com`.

4. **Touch OK to proceed with the purchase.**

   You can also touch Cancel to back out of the purchase if you're having second thoughts!

Figure 16-6: If you try to buy music without a Wi-Fi connection, you get this warning.

5. **Touch I accept to agree to Amazon MP3's terms of service.**

This agreement essentially states that you'll use the downloaded music legally — you agree that it's for private, personal use, and you won't redistribute it.

The application now shows you what it's doing as it attempts to complete your purchase, going through four steps: Purchase started, User authenticated, Credit card charged, and Queued downloads. As each step is completed, a green check box appears to the left of the item. Once all four steps are complete, you see the Downloads screen, which I describe in detail in the next section.

If you're not connected to Wi-Fi, you are once again notified that the downloads can't currently be completed, and you are given an opportunity to connect right now (see Figure 16-7). If you choose not to, your downloads remain queued until you have an opportunity to connect.

Figure 16-7: You can enable Wi-Fi now or later to complete the download.

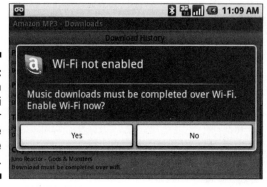

Previewing and buying individual songs

Rather than buying the entire *Gods & Monsters* album, say you're interested in buying only a track or two. By touching an album's row (either while browsing or searching), you see a details screen for that album, as shown in Figure 16-8.

At the top of the screen is a green section showing the album's detail. On the left, the album's artwork appears. To the right, you see the full album name, followed by the artist, the record label, and the average rating (on a scale of one to five stars) given to the album by Amazon reviewers.

Below the album details section is a scrollable listing of the album's tracks. On the left side of each row are the track number and track duration; in the center is the track name; and to the right is a price button that works the same way as the one when purchasing an entire album. Just touch the price button to start the purchase process for that individual track.

You can still buy the entire album from this screen by simply pressing the button with the price on it in the top green section. Let's say you need to hear a track or two before making a decision, though. To hear a brief preview of a track, touch that track's row in the track listing, and the left side of that row changes to red (see Figure 16-9). You see the amount of time remaining in the preview — typically it'll count down from 30 seconds — and a stop icon. Just touch the stop icon, or anywhere else in the row, to stop the preview from playing.

Just like the Music application, you can hear previews over the G1's loudspeaker or your headphones if you have them connected. The phone's Volume buttons control the volume of the preview.

Track duration

Track number

Track name   Album details

Figure 16-8:
The details
screen for
an individual
album.

Managing Music Downloads

The Downloads screen is the clearinghouse for all the tracks you've purchased and downloaded, are currently downloading, or have yet to download. When you purchase new music, you are taken to the Downloads screen
automatically. You can also get to the screen yourself from any screen in
the Amazon MP3 application by pushing the Menu button and selecting the
Downloads menu item.

Continuing with the example in the last section, say that you listened to a few
tracks on *Gods & Monsters* and decided to purchase the entire album without
enabling Wi-Fi on the G1. The Downloads screen indicates that it can't proceed with the download until you hook up to Wi-Fi, as shown in Figure 16-10.

Once you enable Wi-Fi (more on that in Chapter 15), your downloads will begin automatically. The Downloads screen shows you the track that's currently downloading (see Figure 16-11) in a Now Downloading section at the top, along with that track's album artwork, artist name, and a progress bar that indicates how much more data needs to be downloaded for that particular track. Below that, you see how many songs still need to be downloaded after the current track finishes downloading. At the bottom, you see the Queue section — the list of tracks that remain to be downloaded. Only one track is downloaded at a time.

Figure 16-11:
An active download and a queue of tracks waiting to be downloaded.

While tracks are being downloaded, you also get a message in Android's notification system (see Figure 16-12). You can even see your progress from here — when the bar fills up with yellow, all pending downloads have completed. Touching the notification will take you straight to the Downloads screen.

Occasionally, downloads may fail for various reasons. Amazon's servers could be momentarily unavailable, you could have a problem with your Wi-Fi connection, or gremlins could overrun the Internet. When this happens, you see an X symbol to the right of the failed track. Whatever the reason for the failure, you can retry any failed downloads by pushing the Menu button from the Downloads screen and selecting Retry Failed.

As tracks finish downloading, green check marks appear to their right. To clear the list of completed downloads, push Menu and select Clear Completed.

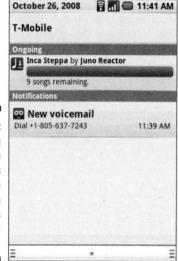

Figure 16-12:
An ongoing
notification
appears
while
Amazon
MP3 is
download-
ing tracks.

Listening to Your Purchased Music

As soon as a track finishes downloading, it goes into your microSD card and is immediately available for listening. Don't believe me? Seriously, check it out — just go to the Music application and your new tracks have magically appeared! As I mention in Chapter 13, the easiest way to find your new music (especially if you have a ton of music already loaded) is to use the Recently added playlist to quickly find new tracks.

Of course, now that you own these wonderful new tracks, you'd probably like to be able enjoy them on your PC, too. In Chapter 17, I talk about the process for transferring them.

Amazon MP3 Settings Screen

From the Amazon MP3 Home screen, push Menu and select the Settings item. There's not much to set here. You can change the Amazon account e-mail address used for making purchases. And if you want to enable 1-Click ordering so you don't have to enter your password each time you make a purchase, select the Enable 1-Click ordering check box by touching it.

Chapter 17

Filling Your Phone with Media

· ·

In This Chapter

▶ Installing and removing memory cards from the G1

▶ Connecting your G1 to your computer

▶ Transferring files between your computer and your G1's memory card

· ·

*U*nlike most smartphones, you could go years without even thinking about connecting your G1 to your PC. Why would you? Your calendar and contacts are being automatically synchronized with Google's servers using your wireless connection, so you know that your most important information is safe and accessible. What's more, that information is accessible from not just your PC but any PC in the world where you can get access to google.com — so why bother ever pulling out a cable and hooking up to your G1 the old-fashioned way?

Well, it turns out there's a very good reason to connect from time to time: when you want to transfer music and pictures. You *could* go a lifetime using Amazon MP3 for all your G1's music needs and using the built-in camera for all your photos, but odds are you probably have some stuff on your computer — your complete collection of live Michael Bolton bootlegs, for example — that you'd just love to have with you in your pocket. With up to 16GB or more of storage space on board, there's no reason not to use the G1 as your full-time portable music player (what's an iPod?) and your photo album on the go.

When it comes to transferring large multimedia files, nothing beats a physical cable connection, and the G1's ExtUSB port offers just the speed you need to quickly move your tunes, images, and other files back and forth. In this chapter, I walk you through the process of hooking the G1 up to your computer. But before that, I show you how to install the microSD memory you need to make the G1 capacious enough for all your files.

Installing and Removing the Memory Card

Your G1 comes with a 1GB microSD card (see Figure 17-1). Although 1GB may sound like a lot, it isn't if you intend to heavily use the phone as your music player (most dedicated music players, such as the iPod, come with at least 4GB of storage these days). As I mention in Chapter 13, cards going all the way up to 16GB in capacity are readily available and surprisingly inexpensive.

Once you've bought a card, installing it is a snap (okay, it's more of a click). The first step is to remove your existing card, if one is installed.

When you remove a microSD card from your G1, the applications you've installed remain installed, but files that you've downloaded or transferred from your PC to your G1 (such as music) go with the card. If you have any files on the card that you'd like to keep, be sure that either you have a way to later read the microSD card directly from your PC (using an SD reader and an SD-to-microSD adapter, for example) or you transfer the files from your G1 to your PC before you remove the card. You can then transfer those files back to the G1 once your new card is installed.

Figure 17-1:
A microSD
card.

To remove a microSD card:

1. **Open the Settings application.**

   You can select its icon from the Applications tab, or you can push Menu from the Home screen and select the Settings menu item.

2. **Touch the SD card & phone storage item.**

   You have to scroll down the screen to see this item.

3. **Touch Unmount/eject SD card.**

   The G1 stops reading from and writing to the card so no data can be damaged when you physically remove it. Alternatively, you can just power off the phone.

4. **Open the keyboard.**

5. **Using your fingernail, pull outward on the small tab to the right of the rightmost shift key on the keyboard (see Figure 17-2) until the tab comes out.**

   Note that the tab and the cover it's attached to remain connected to the phone — you just want the cover to be dangling so it reveals the slot underneath.

Figure 17-2:
The tab on the G1's microSD slot cover.

microSD slot cover   Pull outward

6. **Gently press the card *in* the slot until you hear and feel a click.**

   The card will pop out partway (that's the magic of spring loading). Pull the card the rest of the way out of the phone.

Now that you've removed that dinky little 1GB card, it's time to install a new one. Don't close that slot cover just yet!

1. **Guide your new microSD card into the slot.**

   You can perform this step whether the G1 is on or off. The gold pins on the card should be facing down and forward. Don't force the card into the slot. If you feel a lot of resistance, you're probably inserting the card incorrectly. The card and slot are both fairly fragile.

2. **Push the card until you feel and hear a click (see Figure 17-3).**

The card is now installed and ready to use. You don't need to do anything else; the G1 automatically takes advantage of the new capacity that's available to it.

Figure 17-3:
Carefully push the card into the slot.

Connecting the G1 to Your Computer

Unlike some phones, you don't need any fancy software, drivers, or a call to your technologically inclined nephew to connect your G1 to your computer. In fact, when the two are connected, your PC treats the G1 the same way it treats those USB memory sticks that you might buy at office supply stores — the G1 is just a new "drive" that you can drag files to and from. This feature makes transferring files with the G1 convenient, easy, and fast — just copy your stuff and go.

To connect, you'll need the USB cable that came with your G1. If you misplaced the cable or don't have it handy, you can also use any USB cable that features a mini USB connector on one end and a USB Type A connector on the other. These cables are common for connecting phones and digital cameras, so you might have one or two lying around. If you don't, they're easy to find at any electronics store.

After you've located a cable, make sure the G1 is on, and then do the following:

1. **Connect the mini USB end of the cable (the small connector) to the socket on the bottom of your G1.**

2. **Connect the USB Type A end of the cable (the larger connector) to any available USB socket on your computer.**

   A new symbol appears on the left side of the status bar. You may recognize this symbol as the industry-standard USB logo. Your G1 recognizes that it's been connected!

3. **Pull down the notifications screen on the G1.**

   You see a new notification informing you that your G1's USB connection has been established (see Figure 17-4).

4. **Touch the USB connected notification.**

   A window appears asking you whether you want to mount the memory card to your computer, which allows you to copy files from the card to the computer and vice versa (see Figure 17-5).

While the memory card is mounted to the computer, you won't be able to access the card from your G1 — that is, you won't be able to use the Music or Pictures applications, save any attachments from e-mail or files from the Web, and so on. Only one device can access the memory card at any given time, and when you choose to mount the card to your computer, *only* your computer can access it.

Figure 17-4:
The G1
notifies you
when
its USB
connection
is active.

Figure 17-5:
Confirm that
you want
to mount
the memory
card to your
computer.

5. Touch Mount.

A new drive appears on your computer. Look for the new drive in My
Computer under Windows and on your desktop if you have a Mac. In
either case, double-click the new drive to see its contents.

At this point, you can simply drag files back and forth between your PC and your G1 — treat the new drive just as you would any other drive connected to your computer.

When you open the drive in a window on your computer, you may see several folders that were already created in your G1 memory card:

- ✔ **amazonmp3:** The Amazon MP3 application automatically places any music that you purchase in the amazonmp3 folder. If you double-click the folder to look at its contents, you'll notice that there's one folder for each album (or partial album) you've bought.

- ✔ **dcim:** Pictures taken with your G1's camera go inside a folder called Camera, which resides inside the dcim folder. If you want to grab some pictures that you've snapped and store them in iPhoto, Picasa, or another photo program on your PC, look inside dcim.

- ✔ **download:** Files that you download with the Browser are stored here.

Although certain programs create and use certain folders on your memory card, Android doesn't need them — they're there just for the sake of organization. If you want, you can drag music files directly to the root of the memory card or any folder of your choosing and the Music application will be able to find them. Likewise, you can place pictures wherever you like.

When you have finished transferring files, be sure to "eject" your G1 before unplugging it! This is an important step to making sure that your files are safely and completely written to the memory card. To eject the G1 under Windows, go to My Computer, right-click the G1's drive, and click Eject. On a Mac, simply drag the drive from the desktop to the trash. (The trash icon changes to an eject icon while you're dragging.)

Chapter 18

The Android Market

*B*uying applications for your PC can end up being quite a process. You probably have to run down to your friendly local office supply or electronics superstore, look for the software, and after tearing 15 boxes off the shelves, come to the realization that the store doesn't stock the program (or the version of the program) you need. Next, you hop on the Web and mail-order the program from an online retailer, at which point you're in for at least a day of waiting — and that's if you sprang for the exorbitant overnight shipping option.

There has to be a better way, right? With Android, there is. Google has created a full-featured software store and placed it right on your G1, so you never need to leave the comfort of your handheld to research and download the new application you want. You can be at home, at work, or waiting for a delayed flight in Chicago O'Hare's Terminal 1: The Market is always ready and waiting for you as long as you have a cellular or Wi-Fi connection.

Anyone who's familiar with Apple's App Store for the iPhone will feel right at home with the Market, and you could argue that the Android Market is even better because Apple reserves the right to selectively reject applications that it feels encroach on its own apps. (For example, you'll never see another Web browser offered for sale in the App Store — the iPhone already comes with Mobile Safari, and that's what Apple wants you to use.) Google places no such restrictions, leaving the gates wide open to any and all developers who want to write software and offer it to the Android-using public. Choice is (almost) always a good thing, I say.

Get the credit card ready: Free apps available now, but paid apps coming soon

At the time of this writing, Google had just launched the Android Market (and the G1 and Android itself, for that matter). At first, only free applications are being offered — Google doesn't have the plumbing in place just yet to support automatic billing and payment for developers offering their wares through the system. Even so, though, tons of cool, free apps are already available, and Google is committed to offering paid applications soon. What's more, you'll be able to download free trials of some of these paid applications, making it a cinch to try before you buy.

Out of the box, the G1 comes with almost all the software you need — or at least that's what you might think until you start exploring everything that's available! With endless games, media players, utilities, and informative applets at your disposal, there's no reason *not* to give the Market a spin. In this chapter, I show you how to take advantage of the Android Market to make it your one-stop shop for Android applications.

Starting Market and Understanding the Market Home Screen

To start the Market, select the Market icon from the Applications tab, which is cleverly designed to look like a shopping bag with the Android logo on it. You are taken immediately to the Market home screen (shown in Figure 18-1).

If you've had an opportunity to play with the G1's YouTube application, the Market home screen may seem somewhat familiar. Horizontally across the top, you have access to Featured applications, which are applications that Google has highlighted because they're particularly good or popular or both. You can scroll through the list of Featured applications by touching your finger to the Featured area of the screen and swiping to the left or right. As you scroll, the application in the middle of the screen becomes highlighted and is available to download. Directly below the highlighted icon, the application's name and publisher appear.

Below the Featured area, four items take you to different parts of the Market application: Applications, Games, Search, and My downloads. I cover each of these items in detail throughout the chapter.

Highlighted application

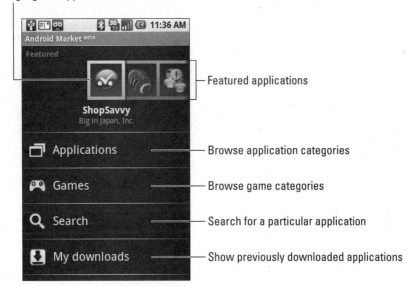

Featured applications

Browse application categories

Browse game categories

Search for a particular application

Show previously downloaded applications

Figure 18-1:
The Market
home
screen.

By pushing Menu from the Market home screen, you see three menu items
at the bottom of the screen. Two of these, Search and My downloads, are
duplicates of the items on the Market home screen itself. The third, Help,
opens the Browser and takes you to the Help section of the Android Market's
official Web site (see Figure 18-2) where you can find out everything there
is to know about downloading applications, managing them once they're
downloaded, and finding solutions to common Market problems. (By the time
you're finished with this chapter, though, I hope you never need to even
think of glancing at Help!)

Figure 18-2:
Help is just
a touch
away when
you're in
the Android
Market.

If you were looking at a Web site in the Browser when you touched the Help option, you'll notice that the Browser shows some courtesy. Instead of unceremoniously navigating away from whatever site you had open so that it could open Android Market Help in its place, it opens Help in a new Browser window; that way, whatever sites you had open remain open. It's a nice touch.

Browsing the Goods: The Applications and Games Screens

Have you ever wandered into a store not needing anything in particular? You want to see what's available, and if anything catches your eye and strikes your fancy, you might just pick it up. (Heck, who knows, maybe that's how you ended up with this book!) I know I have, and the Android Market makes it all too simple.

Touching the Applications or Games items on the Market home screen bring you to a new screen with a list of categories. (See Figure 18-3 for the list of categories for the Applications item.) From here, you can narrow your browsing to a specific genre — say, the News & Weather category within Applications or the Cards & Casino category within Games — simply by touching the category you want. You can also view all available applications or games by touching the All applications or All games items, respectively. (They're very, very long lists, and they're getting longer every day as developers add new applications, so don't say I didn't warn you!)

Figure 18-3:
The
Applications
screen.
Look at
all the
available
categories!

Once you've selected a category, you are brought to a list of applications within it (see Figure 18-4). For this example, let's take a look at the News & Weather category in Applications.

Even with a specific category, as we have here, the list is large. For this reason, the top of the screen has two tabs — By popularity and By date — that help you sort the list in a meaningful way. Touching By popularity (which is selected by default) orders the applications by how frequently they've been downloaded, with the most frequently downloaded at the top. Touching By date presents applications according to when they were added to the Market, with the newest at the top of the list.

If you want a quick peek in the Market to see if anything is new and exciting since your last visit, go to the Applications or Games screen, choose All applications or All games accordingly, and touch the By date tab. You see new apps at the top of the list. If you visit frequently, you can simply read down the list until you see an application you recognize to get a picture of everything that's been added since the last time you were here.

Orders applications by download frequency

Orders applications by when they were added to the Market

News & Weather

By popularity   By date

| The Weather Channel | FREE |
| The Weather Channel | ★★★★✦ |

| AccuWeather.com | FREE |
| Accuweather.com | ★★★★ |

| WeatherBug (Beta) | FREE |
| WeatherBug Mobile | ★★★✦ |

| iMap Weather™ | FREE |
| Weather Decision Technolo... | ★★★✦ | — User rating

| Weather | FREE |
| Michael Bachman | ★★★ |

| VOTE Vote2008 | FREE |

Figure 18-4: The list of applications available in the News & Weather category.

Application icon   Application publisher   Application name

Below the tabs, each row represents an application you can download. To the far left is the application's icon; this is a slightly smaller version of the icon you see in the Applications tab when the new application is installed. To the right of the icon is the application's name and the name of its publisher (that is, the name of the company that produced the software). At the far right, you see the price of the application (everything is labeled FREE until Google introduces its billing system to the Market) and the application's average user rating on a scale of 1 to 5 stars. (If you download an application and want to add your own rating, hang tight — I get to that topic shortly.)

Getting Details about and Installing Applications

When you select an application in the list in the Market, you see a details screen for that app, as shown in Figure 18-5. In the details screen, you can find out pretty much everything you need to know to make a decision about whether you want to download and install the application. (If you still can't decide, don't hesitate to download the application and give it a try anyway — odds are it's either free or available in a free trial version.)

As you can likely tell at a glance, the details screen has a lot of information. Let's take a look, starting from the top:

Figure 18-5:
An application's details
screen.

✔ **Application icon, name, publisher, price, and user rating:** This is the same information you saw in the list of applications before you selected one of them and were brought to this screen. These key pieces of information are summarized in a gray bar at the top of the screen, and as you scroll through the remainder of the information, this part of the screen remains fixed so you can always see it.

✔ **Number of downloads and ratings:** Near the top of the screen, you see a rough estimate of the number of times this application has been downloaded (somewhere between 10,000 and 50,000 for this AccuWeather. com app, for example) and the number of users who've rated it on the 1- to 5-star scale.

✔ **Description:** Next, you see a brief description of what this application is and what it does. Sometimes, an applications description is unnecessary (AccuWeather.com probably falls into that category), but other times the application is a mystery until you read this section.

✔ **Version and size:** Publishers may periodically update their applications in the Market to fix bugs and add features. As they do, the Version, which is located directly below the application description in smaller gray text, increments to progressively higher numbers.

You'll typically find these Versions in the format X.Y or X.Y.Z, where X, Y, and Z are all numbers (AccuWeather.com, for example, is currently at Version 1.0.0). A change in X represents a more significant change in the application than a change in Y; likewise, a change in Y is a more significant change than a change in Z. It's common for developers to change only Y or Z when fixing bugs and adding minor features, and to change X when adding major new functionality or revamping the application.

The size of the application is indicated just to the right of the Version in kilobytes (KB) or megabytes (MB).

Currently, applications can be stored only in the G1's internal memory, of which only 70 megabytes or so is available to you. I hope a future version of Android will allow you to store applications on your much more expansive microSD card as well, but until that happens, pay close attention to the size of each application you're installing and monitor which applications you actually use and which are just stagnating in the Applications tab. Depending on how extensively you use the Market, you may find that you have to pick and choose carefully!

✔ **Comments:** The next section in the application's details screen lists the most recent comments left by users who've downloaded it. On the top left of each comment row is the name of the user who left the comment and the date it was written. To the right, you see that user's 1- to 5-star rating. Below all that, you see a snippet of the comment.

Users can be vocal, to say the least — and only a small handful of comments are shown on the application's details screen itself. To see more, touch the Read all comments row at the bottom of the Comments section. You are taken to a new screen that displays every comment (in its entirety, not just the snippet you see on the details screen). Once you have your fill of independent opinions, push the Back button to return to the details screen.

✔ **About the developer:** At the bottom of the screen, you're presented with two items in the About the developer section. Touching View more applications presents a list of all applications in the Market that were published by the same developer — a good thing to check if you are happy with an application and are curious what else the developer has created.

Have praise, complaints, suggestions, or bugs to report? The second item, Send email to developer, opens a window that allows you to choose between the Email and Gmail applications for sending an e-mail to the application's developer. Their e-mail address is filled in for you in the To field and the e-mail's Subject line contains the application's name — just whip up the body of the message, touch Send, and you're all set.

✔ **Install, Open, and Uninstall:** If the details screen you're looking at is for an application that isn't installed on your G1, you see a single button at the very bottom of the screen: Install. If the application *is* installed, however, you see two buttons labeled Open and Uninstall instead. Touching Open simply starts the application, as if you had touched its icon in the Home screen's Applications tab. Touching Uninstall does the same thing as uninstalling an application from Android's Settings. If you touch the Uninstall button, you see a window prompting you to confirm that you want to remove the application from the phone (see Figure 18-6); touch OK to proceed. As the window says, you can always use the Market to install the application again later if you have a change of heart.

When you're ready to install an application, just touch the Install button from application's details screen.

Before installation proceeds, you may be shown a list of things the application can do with your G1 and what data of yours it will have access to. Read this list carefully and make sure you're comfortable with it! Items of particular interest are colored in yellow and shown with an exclamation point to their left; other less-risky items are automatically collapsed under a heading at the bottom labeled Show all (just touch Show all to see the full list). In Figure 18-7, for example, I'm trying to install iMap Weather. Android lets me know that iMap Weather will have access to my Internet connection and my current location, and will be able to modify system settings. Generally speaking, you don't need to worry, because most applications aren't trying to harm you or your G1. But just like your PC, malicious applications can exist for your phone, and you should be on the lookout for anything in this list that seems out of place. For example, if you were to encounter a simple video player that wanted access to your contacts, that might raise a red flag that the application might collect e-mail addresses to spam them.

Figure 18-6: A confirmation window gets your approval before uninstalling an application.

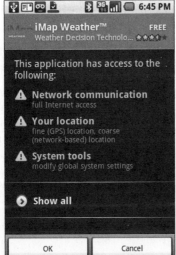

Figure 18-7: Android warns you of capabilities and data that this application will have access to once you install it.

Google is constantly monitoring the Market to remove "bad" applications as quickly as possible. As a last resort, Google has the capability to remotely disable applications on your phone that it discovers are harmful. Furthermore, all applications that are made available in the Market must be *digitally signed*, a technique that guarantees the source of the application and prevents applications from being offered anonymously. That being said, it's always good to take a quick glance at this list before moving forward with the installation — better safe than sorry!

If you're comfortable with everything that this application will have access to, touch OK to continue. To back out of the installation, touch Cancel.

When you touch OK, the application begins downloading (you can be connected to a cellular or a Wi-Fi network for this, but keep in mind that a Wi-Fi network will always be faster). As soon as each application you download has finished transferring, Android automatically installs it so it's ready for you to use.

The Market reminds you with a message on the screen that you should keep an eye on the notifications in your status bar for the progress of your download. That's solid advice, because the status bar is where you find out what's currently downloading and what has finished installing (see Figure 18-8). At this point, you can keep browsing applications or exit the Market entirely — your applications will continue to quietly download and install in the background without any more intervention from you.

When installation is complete, the application's icon appears in the Applications tab. You can also just pull down the notifications screen and touch the name of the newly installed application to start it. Congratulations — you've successfully installed and launched your first application from the Market!

Searching for Applications

Sometimes you're not in the mood to browse for new wares for your G1 — you know exactly what you want; you just have to *find* it. No sweat. From anywhere in the Market, you can search for available applications like so:

1. **Push the Menu button and then select the Search menu item.**

   Or if you have the keyboard open, you can press the Search key or just start typing your search — the search bar will automatically open. You can also search for an application from the Market home screen by touching the Search row (refer to Figure 18-1).

2. **Type your search into the search bar that appears.**

   Your search can be composed of an application name, a publisher name, or a category. As you type, a list of previous searches appears below the search bar. If you see what you're looking for in the list, just touch the item (or select it with the trackball); your search will begin and you can skip Step 3.

3. **When you've finished typing your search terms, press the Enter key or touch the Search button.**

   After a moment of searching, the Market returns a list of applications that match your search.

4. **In the list, touch the item you're looking for.**

Completed downloads

Download in progress

Figure 18-8: Downloading and installed applications are indicated in the status bar.

The My Downloads Screen

After you've downloaded an application for the first time, it automatically moves to the My downloads screen (see Figure 18-9). To access this screen, touch My downloads on the Market home screen (refer to Figure 18-1), or push the Menu button from anywhere else in the Market and select the My downloads menu item.

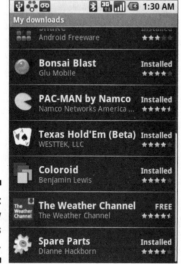

Figure 18-9:
The My
downloads
screen.

Once you download an application, it remains in the My downloads list even if you uninstall it. In Figure 18-9, for example, notice that every row says Installed on the right side except for The Weather Channel — that's because I had The Weather Channel installed at one time, but I've since uninstalled it. If you're uninstalling rarely used applications to conserve your G1's limited internal memory, the My downloads screen makes it especially easy to reinstall them at a future time should the need arise.

Canceling downloads in progress

Applications appear in My downloads the moment that you start downloading them, which gives you the opportunity to cancel a download if you change your mind:

1. **Go to the My downloads screen.**

   While an application is downloading, it appears in a special Currently downloading section of the My downloads screen that appears at the very top. The progress of each download is indicated as a yellow bar that moves across the screen.

2. **Touch and hold on the application whose download you want to cancel.**

   After a moment of holding down, a pop-up menu appears.

3. **Select Cancel download.**

Help out fellow G1 owners by rating applications

After you've used applications for a little while, you probably will have developed some opinions. Do they work well? Do they do what they said they would? Do they have any annoying bugs? And most importantly — are they worth the download?

Rating the applications you've downloaded on a 5-star scale is a piece of cake, and an easy way to convey to owners of Android devices how good (or how awful) a particular application is. Here's how to do it:

1. **Go to the My downloads screen.**

2. **Touch the row of the application you want to rate.**

   The application's details screen appears, but it has a key addition near the top of the screen that you won't have seen before you down-loaded the application: a My review section.

3. **Touch Rate it!**

   The window in Figure 18-10 appears, where you can indicate how many stars you want to give this application — 5 means "totally awesome," and 1 means "I can't believe I wasted my time downloading this hunk of garbage." (I'm paraphrasing here, but you catch my drift.)

Figure 18-10: Rating an application that you've down-loaded.

4. **Touch a star on the screen.**

   The leftmost star equates to a rating of one star and the rightmost equates to a rating of five stars.

5. **Touch OK to confirm your rating or Cancel to back out without saving it.**

After you've rated the software, a new item, Post a comment, appears in the My review section of the screen directly below My rating. By touching Post a comment, a text box appears where you can wax poetic about the highs and lows of this application. As with the rating, touch OK to confirm or Cancel to discard your comment.

Part V
The Part of Tens

The 5th Wave   By Rich Tennant

"This model comes with a particularly useful function — a simulated static button for breaking out of long-winded conversations."

In this part . . .

These days, Android is surrounded by a ton of attention and buzz — and with that buzz comes a whole bunch of accessories, applications, and miscellany as developers and manufacturers try to ride on the G1's successful coattails. How's a new G1 owner to separate the wheat from the chaff? Join me in Part V as I look at lists of my favorite applications, accessories, Android-related Web sites, and games.

Chapter 19

Ten Great Android Applications

*I*f you've taken a glance at Chapter 18, you're probably well acquainted with the Android Market. There's a problem, though: With hundreds of applications offered in the Market and more added every day, it can be nearly impossible to cut through the fat and figure out what's worth downloading.

That's a tough problem to solve — and no two people will love exactly the same Android applications — but I've boiled down the treasure trove of goodies offered by the Market to a list of ten must-try apps. Go ahead and live it up. The applications are free, after all, and if you decide an app isn't for you, you can always uninstall it.

CompareEverywhere

It turns out that your ultimate shopping companion *isn't* a pocketbook full of cash and plastic (although that certainly helps). With CompareEverywhere, just scan the UPC codes found on products you're shopping for, and within a few seconds, the app will tell you if another nearby store has a better price and what you'd pay if you were to mail order the same product from an online retailer. You'd be amazed by how much you can save.

TuneWiki

The G1's built-in music player isn't the most incredible music player I've ever used, and it probably isn't the best you've ever used, either. It gets the job done, but better apps are out there to blast your tracks, and TuneWiki is one of them. With TuneWiki, you can see synchronized lyrics (think karaoke in your pocket) for many of the songs in your library, chat about tracks, artists, and playlists with other TuneWiki users, and even see the location of other users and what they're listening to.

Locale

So-called location-based services (you may hear the acronym *LBS* thrown around from time to time) are all the rage on phones these days, and Locale is a perfect example of what makes them so incredible. The application allows you to automatically configure certain Android settings — your network setup, your ringtone, and even your Home screen wallpaper — depending on your location, remaining battery life, current date and time, and who's calling. If you're going to download just one application for your G1, this should be it!

Caller ID by WhitePages

One popular feature of the Apple iPhone is its ability to identify the city and state of incoming calls — a handy tool when a phone number is unfamiliar. CallerID by WhitePages brings that same functionality to Android and takes it a few steps further by automatically looking up incoming numbers in public records, displaying any names associated with them, and letting you know whether calls are coming from cellphones or landlines.

Shazam

What CompareEverywhere is to products on store shelves, Shazam is to the music you hear. This ultra-handy little app listens to tunes, identifies what they are, and gives you the opportunity to buy the tracks immediately. If you just want to store track names for later, that's cool too because Shazam keeps a list of every song it's identified. It's amazingly accurate.

Video Player

For some strange reason, Google didn't bother to include a video player on the G1 beyond its own YouTube application. Don't get me wrong; YouTube's great. But sometimes you want to watch your *own* video, not what YouTube has to offer. For that, you'll want one of the video player applications available in the Market, and the aptly named Video Player is about as simple and straightforward as they come. Just drop MPEG4 and 3GPP video files on your G1's memory card, and Video Player will eat 'em up.

Any Cut

Out of the box, Android lets you add shortcuts to contacts, Gmail labels, music playlists, and Browser bookmarks right onto your Home screen. Any Cut goes one step further by adding support for Home screen shortcuts to direct calls and text messages as well. Translation: speed dial (and speed text message) right from your Home screen. Awesome.

The Weather Channel

What's the single most important piece of information you want when you wake up in the morning? (And no, a cup of coffee doesn't count as a "piece of information.") I like to see the weather forecast for the day so I can plan what I'm wearing and figure out whether I need to lug around an umbrella. No shortage of weather applications are available for Android, but The Weather Channel is the slickest I've seen, with an easy-to-use tabbed interface and the ability to spit out the forecast for your current location automatically.

OI Notepad

You already know that Google forgot to include an all-purpose video player with the G1. You know what else they forgot? A note-taking app! OI Notepad is nothing more than an ultrabasic way to jot down quick notes — it's not fancy, but it gets the job done.

Ringdroid

The G1's Music application lets you assign a track as your ringtone, but it assigns the whole darned song. Sometimes the beginning of a track isn't the best part to turn into a ringtone. Ringdroid lets you cut small sections out of your stored music and use the snippets as ringtones. What's more, you can record new ringtones using the G1's microphone.

Chapter 20

Ten Accessories for Your G1

*Y*ou didn't think that your spending spree would end the moment you picked up your G1, did you? Like a pet, a car, or an expensive (but enjoyable) hobby, you can more or less spend as much money accessorizing your new handset as you please. In this chapter, I take you through a quick list of the things you might want to look at buying first.

Bluetooth Headset

In many states and municipalities, talking on a handset while driving is now a primary offense, meaning you can be pulled over for it and ticketed (and yes, your insurance rates might go up, too). Because of that, there's never been a better time to invest in a Bluetooth headset that you can use to help you keep your hands on the wheel while you're on the road.

Bluetooth headsets are wildly popular these days, so they are easy to find; are available in every shape, style, and color of the rainbow; and range in price from just a few dollars to several hundred. One of my favorite models for style, noise resistance, and comfort on the ear is the Jawbone, available from www.jawbone.com for $129.99 at the time of this writing. However, you can get a perfectly good headset for much, much less by visiting any wireless store or your T-Mobile location.

microSD Card

I've mentioned a few times throughout the book that the 1GB microSD card included with the G1 isn't enough to maximize the phone's capabilities. To use the G1 as an effective music and photo player, you'll want more memory. At Best Buy, amazon.com, and virtually any other retailer that sells electronics, you can find 16GB cards for well under $100 and 8GB cards for $20 or less!

microSD-to-SD Card Adapter

The microSD standard is a miniaturized version of the SD standard, a popular memory card format used by a variety of devices including most consumer digital cameras and the Nintendo Wii. Being able to move your memory cards among all your devices is convenient, and a microSD-to-SD card adapter allows you to do that. These adapters look like SD cards and contain a slot on one end for inserting a microSD card. They're sometimes included with microSD cards you buy, but if you don't have one, your best bet is to get one bundled with a 512MB or 1GB microSD card for under $10.

USB SD Card Reader

Another benefit of having a microSD-to-SD card adapter handy is that when you add a USB SD card reader into the mix, you'll be able to pull your memory card out of your G1, plug the card right into your PC, and read it (and write to it) without having to connect the G1 itself. These readers plug into an available USB port on your PC and have an SD slot on the other end; just plug in a card and you're off to the races. A great place to find a wide selection of memory card readers, many of which sell for just $6 or $7, is www.newegg.com.

Spare Battery

The G1 is a powerful phone, and an ancient saying applies here: "With great power comes great power drain." If you frequently have a laptop or messenger bag with you, it's not a bad idea to keep a charged, ready-to-go spare G1 battery in there in case you've been using the phone heavily and you're at risk of not making it back to a charger in time. At the time of this writing, T-Mobile doesn't sell spare G1 batteries, but I expect them to start doing so soon for between $40 and $50.

Screen Protector

All phone screens are at constant risk of nicks, cracks, and scratches, but the G1 is at even more risk than your average phone because you're constantly touching and interacting with the screen. You can find a number of sources for phone and PDA screen protectors on the Internet. Fellowes (www.fellowes.com) produces the WriteRight series, which has been around for many years and has always served me well. Finding WriteRights or any other brand of screen protection presized for your G1 may be difficult, but they are available in a universal size that you can easily trim to fit the G1's screen.

ExtUSB-to-3.5mm Jack Adapter

If you're looking to use the G1 as a serious music player, one of the first things you need to do is ditch the included headphones. But to do that, you need an adapter that allows the phone to accept regular headphones that use a 3.5mm jack, and not all G1s include such an adapter. One of the nicest aftermarket adapters that I've seen is the 3-in-1 Adaptor from eXpansys (www.expansys-usa.com), retailing for about $13 and taking up as little room in your G1 (or in your bag) as possible.

Upgraded Headphones or Earphones

Now that you've nabbed that ExtUSB-to-3.5mm jack adapter, the next step is to get something to plug into it. You'll have no problem finding decent headphones or earphones from any electronics retailer — and seriously, anything is better than what T-Mobile includes with the G1. However, if you're serious about your sound, I recommend checking out Shure (www.shure.com) or Ultimate Ears (www.ultimateears.com). Their extremely high-quality earphones range in price from $100 up to a staggering $1150!

Car Charger

If you're doing any amount of driving, it's handy to be able to use your car as a backup method of charging your G1. Fortunately, the G1 charges from a standard USB connector, so you can use any car charger that features a USB port. Just connect your USB cable to the charger, connect the G1 to the other end, and you'll start the flow of much-needed juice to the battery. USB car chargers are relatively easy to find these days, and you should have no trouble coming across a few choices at your local Best Buy.

Case

As hardy as the G1's shell might be, it will still get banged up over time if you're using it heavily. (And with that great e-mail and Web access, why wouldn't you be using it heavily?) A case can save you from the heartache of watching your beloved phone take the brunt of the abuse. You can find cases in all sorts of styles, everything from form-fitting cases that slip onto the shell without changing its shape to big leather pouches equipped with extra pockets and slots for accessories. T-Mobile has yet to offer any cases directly for the G1, but I suspect they will. If you're in the market for an ultra high-quality leather case, check out Vaja (www.vajacases.com), where you can choose from a variety of styles and colors.

Chapter 21

Ten Great Sites for Android Information

● ●

In This Chapter

▶ Official Android site

▶ T-Mobile Forums

▶ HTC T-Mobile G1 support

▶ Android Forums

▶ HowardForums

▶ Android Community

▶ AndroidGuys

▶ Phandroid

▶ Phone Scoop

▶ Engadget Mobile

● ●

So you've settled in with the G1, you're taking advantage of the Market to get some cool apps installed, and you're all set for accessories. Now, the next challenge is staying on top of things! Android is a young platform that's changing rapidly and almost constantly; major new software versions will be made available and cool new handsets will be announced at a rapid clip. Fortunately, you can use a dedicated array of Web sites to stay one step ahead of all the Android news that's fit to print, and I'll take you through a few of my favorites in this chapter.

Official Android Site

www.android.com

What better way to kick off this list than with Android's official home page on the Web? This is the place to read up on the latest news about Android. The site also features links to Android's developer and Market communities, where you can read questions and answers from forum members and add your own comments into the mix.

T-Mobile Forums

forums.t-mobile.com

T-Mobile's official forums have three active message boards dedicated to the G1 where you can go to find answers to problems users are having and lend a helping hand in solving others' issues. Even better, T-Mobile staffs these forums with employees who occasionally chime in with an official word from the company, so this is definitely the place to go if you want your voice to be heard.

HTC T-Mobile G1 Support

www.htc.com

If you should ever lose the user's manual that came with the G1, you can grab a digital copy from HTC's own support site. To get here, go to HTC's home page at www.htc.com, click the Support tab at the top of the page, and then click T-Mobile G1 in the list that appears. You'll also find a list of frequently-asked questions (FAQs) maintained here.

Android Forums

androidforums.com

The independently run Android Forums site is not monitored (at least not officially, anyway) by any T-Mobile, HTC, or Google staff, but it is active and has a wider array of message boards available than T-Mobile's own site. If you're looking for help from your fellow players in beating a particular game, for example, Android Forums is a great place to start looking.

HowardForums

www.howardforums.com

Widely regarded as the one of the most active communities of phone users on the Internet, HowardForums isn't just for G1 users — or even just Android users, for that matter. Instead, it's a huge melting pot of folks from nearly every carrier across North America (and the globe) looking to chat about phones, phone technology, phone software, and everything in between. You'll find HTC, T-Mobile, and Android message boards here, among many others.

Android Community

www.androidcommunity.com

Android Community is a comprehensive site for all things Android. The site consists of a newsy blog, a database of Android applications, and discussion forums. The blog is a good daily read to keep up on the latest goings-on in the world of Android and the G1.

AndroidGuys

www.androidguys.com

Like Android Community, AndroidGuys seeks to be your one-stop shop for Android information on the Web. It does a solid job with software reviews (which are going to become more and more important as the Market fills to the brim with a nearly endless list of available applications), and the editorials make for an entertaining read.

Phandroid

phandroid.com

Phandroid covers Android news and reviews in a blog format. I wouldn't look to it as my primary source of news about the G1, but the site does a solid job of staying on top of interesting Android applications that might otherwise slip under the radar.

Phone Scoop

www.phonescoop.com

In addition to covering mobile industry news, Phone Scoop is your encyclopedic reference for virtually everything related to cellphones. Android is given equal time here along with every other platform — Windows Mobile, Symbian, and so on. And if you should ever need to know the screen resolution of a Nokia 3250, Phone Scoop is the first, last, and *only* place you should look.

Engadget Mobile

www.engadgetmobile.com

Of course, I'd be remiss if I didn't plug the site that I write for. A dearly loved child of tech blog Engadget, Engadget Mobile covers everything in the mobile business with news, reviews, and editorials published 365 days a year. If you want to be the very first to hear about that new Android handset or the latest firmware update for your G1, Engadget Mobile is the place to go. (And if you just don't want this book to ever end, take comfort in the fact that you can read me there nearly every day of the year.)

Chapter 22

Ten Must-Play Android Games

As far as entertainment goes, the YouTube application will hold you over for only so long before you have to give in and pick up some games. Actually, on second thought, maybe YouTube's endless collection of bizarre and hilarious videos *could* be enough to entertain you forever. But trust me; it's worth your time to explore the lighter side of the Android Market by downloading a few of these fantastic titles.

PAC-MAN by Namco

Vintage video games are making a comeback, and no catalog of arcade classics would be complete without the timeless PAC-MAN. Namco's version for Android is a faithful reproduction of the original, and three control modes make the game surprisingly easy to use.

Bonsai Blast

If you're looking to hone your thumb-eye coordination, Bonsai Blast is just the ticket. Shoot colored spheres into an ever-advancing line to form and destroy combinations before the line reaches the end! Tons of fun levels make Bonsai Blast playable for hours upon hours — just don't be surprised if you end up with a sore thumb.

Texas Hold 'em

Is "all in" a way of life for you? Texas Hold 'em is one of the first games to show up for Android, which is no surprise given the popularity of the card game of the same name. The application looks great and works well. (I especially like the innovative betting mechanism — just rotate a chip to change the bet.) However, in its current incarnation, your only option is to play against a host of computer opponents. A multiplayer version would be even better!

Snake

If you used a Nokia phone at any point in the late 1990's or the early part of the 2000's, you already know what this game is all about! Snake is a remake of the Nokia classic, albeit on a slightly larger screen and with just a bit more color than those tiny monochrome screens of yore. The concept is to guide around the screen an ever-growing snake, which is eating fruits to get longer and move more quickly. If you crash into yourself or the wall, the game is over. Snake is a quick, simple game that you can easily fit into a spare five minutes here and there.

Coloroid

Coloroid is another simple but deceivingly difficult game. You're presented with a board of colored squares, and your goal is to turn it all into a single color in as few moves as possible by making wise choices in expanding the selected area of the board. As your skill level improves, you can increase the number of squares on the board until you're ready to pull out your hair!

Light Racer

Remember the movie *Tron* with those crazy light cycles? You may also remember — and in doing so, you'll be showing your age — that the movie was followed by a Midway arcade game where you raced those light cycles around the screen, trying to cut off your opponent and force him to crash. Light Racer is a remake of the Midway original, and it works really well — just get ready to be quick with the trackball.

Divide and Conquer!

Divide and Conquer! is another classic that you may have played in one of its many forms on a PC or video game console in years past. The idea is simple: You have some balls on the screen, and you must contain them in a small area by drawing lines. If a ball hits a wall while it's still being constructed, you lose a life. It takes a steady hand and a watchful eye to make it past the first few levels.

MisMisMatch

If you are looking for an SAT-style mental workout, look no further than MisMisMatch. The game — which would actually qualify more as schoolwork to some, I think — takes you through a series of flashcards of colored shapes. Your goal is to find the series of three cards that are different in *two* ways. It's harder than it sounds.

Parallel Kingdom

Parallel Kingdom isn't great just yet, but it has the potential to become one of Android's killer games if enough people start to use it. The game uses your actual location to add villains and items to the map of the area surrounding you in real life. The idea is not novel but it is the first time it has been made available in a device as high-profile as the G1. As people around you start to play the game, they are added to your map as well so that you can interact with them, which is where things could start to get interesting. Here's hoping!

Solitaire

In my opinion, no electronic device is complete until it has some form of Solitaire loaded, and the G1 is no exception. The version available in the Android Market works just like the ever-present time waster that you find in every version of Windows, but it also includes Spider and Freecell versions for endless one-player entertainment.

Index